God and the Multiverse

In recent decades, scientific theories have postulated the existence of many universes beyond our own. The details and implications of these theories are hotly contested. Some philosophers argue that these scientific models count against the existence of God. Others, however, argue that if God exists, a multiverse is precisely what we should expect to find. Moreover, these philosophers claim that the idea of a divinely created multiverse can help believers in God respond to certain arguments for atheism. These proposals are, of course, also extremely controversial. This volume collects together twelve newly published essays—two by physicists, and ten by philosophers—that discuss various aspects of this issue. Some of the essays support the idea of a divinely created multiverse; others oppose it. Scientific, philosophical, and theological issues are considered.

Klaas J. Kraay is an associate professor of philosophy at Ryerson University in Toronto, Canada. He has published articles in such journals as *Philosophical Studies, Erkenntnis, American Philosophical Quarterly, Canadian Journal of Philosophy, International Journal for Philosophy of Religion, Religious Studies,* and *Faith and Philosophy.*

Routledge Studies in the Philosophy of Religion

1 **God and Goodness**
A Natural Theological Perspective
Mark Wynn

2 **Divinity and Maximal Greatness**
Daniel Hill

3 **Providence, Evil, and the Openness of God**
William Hasker

4 **Consciousness and the Existence of God**
A Theistic Argument
J.P. Moreland

5 **The Metaphysics of Perfect Beings**
Michael J. Almeida

6 **Theism and Explanation**
Gregory W. Dawes

7 **Metaphysics and God**
Essays in Honor of Eleonore Stump
Edited by Kevin Timpe

8 **Divine Intervention**
Metaphysical and Epistemological Puzzles
Evan Fales

9 **Is Faith in God Reasonable?**
Debates in Philosophy, Science, and Rhetoric
Edited by Corey Miller and Paul Gould

10 **God and the Multiverse**
Scientific, Philosophical, and Theological Perspectives
Edited by Klaas J. Kraay

God and the Multiverse

Scientific, Philosophical, and
Theological Perspectives

Edited by Klaas J. Kraay

Routledge
Taylor & Francis Group

New York London

First published 2015
by Routledge
711 Third Avenue, New York, NY 10017, USA

and by Routledge
2 Park Square, Milton Park, Abingdon, Oxon OX14 4RN

First issued in paperback 2017

*Routledge is an imprint of the Taylor & Francis Group,
an informa business*

© 2015 Taylor & Francis

Library of Congress Cataloging-in-Publication Data

God and the multiverse : scientific, philosophical, and theological perspectives /
    edited by Klaas J. Kraay. — 1 [edition].
        pages cm. — (Routledge studies in the philosophy of religion ; 10)
    Includes bibliographical references and index.
    ISBN 978-1-138-78867-1 (alk. paper)
    1. Cosmology. I. Kraay, Klaas J., 1975– editor.
    BD511.G63 2014
    215'.3—dc23
    2014019871

ISBN 13: 978-1-138-30220-4 (pbk)
ISBN 13: 978-1-138-78867-1 (hbk)

Typeset in Sabon
by Apex CoVantage, LLC

Contents

Introduction   1

Physicists on God and the Multiverse

1   Puzzled by Particularity   25
    ROBERT B. MANN

2   The Everett Multiverse and God   45
    DON N. PAGE

Theistic Multiverses: Details and Applications

3   The Multiverse: Separate Worlds,
    Branching, or Hyperspace? And What Implications
    Are There for Theism?   61
    PETER FORREST

4   An Argument for Modal Realism   92
    JASON L. MEGILL

5   Revisiting the Many-Universes
    Solution to the Problem of Evil   114
    DONALD A. TURNER

Criticisms of Theistic Multiverses

6   Kraay's Theistic Multiverse   129
    MICHAEL SCHRYNEMAKERS

7   Best Worlds and Multiverses   149
    MICHAEL ALMEIDA

8   On Multiverses and Infinite Numbers   162
    JEREMY GWIAZDA

Pantheistic Multiverses

9   Multiverse Pantheism   177
    YUJIN NAGASAWA

10 God and Many Universes   192
    JOHN LESLIE

Multiverses and the Incarnation

11 Extraterrestrial Intelligence and
    the Incarnation   211
    ROBIN COLLINS

12 Incarnation and the Multiverse   227
    TIMOTHY O'CONNOR AND PHILIP WOODWARD

    Contributors   243
    Index   245

Introduction

Since one of the most wondrous and noble questions about nature is whether there is one world or many, a question the human mind desires to understand *per se,* it seems desirable for us to inquire about it.

—Albertus Magnus

Thus is the excellence of God magnified and the greatness of his kingdom made manifest; he is glorified not in one, but in countless suns; not in a single earth, a single world, but in a thousand thousand, I say in an infinity of worlds.

—Giordano Bruno

On March 17, 2014, Stanford University released a remarkable two-minute video.[1] It shows Chao-Lin Kuo, a faculty member in the Department of Physics, arriving unannounced at the suburban home shared by two of his colleagues, Andrei Linde and Renata Kallosh. As Linde and Kallosh open the door, Kuo, without salutation or preamble, announces: "So, I have a surprise for you. It's five sigma at point two". Kallosh, visibly shocked, manages to blurt out one word: "Discovered?" "Yes", Kuo replies. Kallosh immediately embraces Kuo, while Linde, astonished, twice asks him to repeat himself. He can hardly believe what he is hearing. Moments later, the three physicists can be seen raising a champagne toast. As the video concludes, Linde turns to Kuo and, in voice trembling with emotion, says "Thank-you so much for doing this".

Kuo was reporting the latest results from a research project that has been using increasingly sophisticated radio telescopes at the South Pole to examine Cosmic Microwave Background radiation emitted during the infancy of the universe. In particular, the project aims to detect a certain swirly pattern in polarized light known as 'B-mode polarization'.[2] It is widely thought that evidence of this phenomenon would provide the strongest support yet for the theory of cosmic inflation. Linde is one of the pioneers of this theory, which posits that just 10^{-35} seconds after the Big Bang, the universe expanded by one hundred trillion trillion times in less than the blink of an eye.[3] Many believe that this theory strongly suggests the existence of other universes beyond our own. Linde himself puts it simply and directly: "If inflation is there, the multiverse is there".[4] The results have yet to be published, and they may, of course, be disconfirmed. But if they stand up to scrutiny, this may well count as one of the most important discoveries in the history of cosmology. Several prominent scientists are already on record suggesting that the work of Kuo and his colleagues, if confirmed, will merit a Nobel Prize.[5]

The dramatic moment captured in this video represents one of the most recent episodes in a history of scientific, philosophical, and theological inquiry stretching back at least 2,500 years in the Western world.[6] In the fifth century BCE, the Greek atomists Leucippus and Democritus posited the existence of innumerable realms beyond our own, each with an earth at its center, surrounded by planets and stars. In the third century BCE, Epicurus held the same view, and it was popularized by Lucretius in the first century BCE. In contrast, Plato (c. 428–347 BCE) and Aristotle (384–322 BCE) rejected the idea of a plurality of worlds, and some early Christian thinkers such as Hippolytus, and Philastrius, in the third and fourth century CE, respectively, deemed it to be heretical. Augustine (354–430) and Aquinas (1224–1274) rejected it as well.

Three years after Aquinas' death, however, a momentous event occurred. In 1277, the Bishop of Paris condemned 219 propositions as heretical. The thirty-fourth of these asserted that God could not make several worlds. The condemnation of this claim allowed late medieval Christian thinkers to more openly consider the notion of multiple worlds, although they nevertheless generally rejected it.

When the Copernican revolution replaced the geocentric view with the heliocentric view, this paradigm shift and its many attendant astronomical discoveries ushered in an era even more hospitable to the consideration and development of many-worlds theories. Giordano Bruno (1548–1600) vigorously defended the plurality of worlds, as did Christiaan Huygens (1629–1695) and Bernard le Bovier de Fontenelle (1657–1757), both of whom claimed that stars were encircled by other planets, and indeed that these were inhabited. Together with the discoveries of Galileo Galilei (1564–1642) that diminished the overall significance of earth's sun, the heliocentric view was eventually supplanted by what Carr (2007) calls the *galactocentric* view, according to which reality consists in the entire Milky Way.

The eighteenth and nineteenth centuries saw enormous debate among the leading intellects about the existence, nature, and scope of other worlds, whether they might be inhabited, and whether and to what extent they might conflict with Christian doctrine. Immanuel Kant (1724–1804) was one of the first to speculate that the Milky Way might not exhaust all of reality.

These speculations were confirmed by Edwin Hubble (1889–1953) in 1924, thus ushering a shift to what Carr (2007) calls the *cosmocentric* view.

In the middle of the twentieth century, Milton Munitz (1951, 254) wrote that "the essential problems confronting cosmology at the present time do not include active debate as to whether there is more than one universe". Since then, however, multiple universe theories have proliferated, and they have increasingly gained scientific respectability. Today, there are many theories of multiple universes under serious consideration by physicists. As a result, Carr (2007) deems the cosmocentric view to have been replaced with the *multiverse* view. Some of the more prominent theories include (1) Everett's (1957) *many-worlds* interpretation of quantum mechanics, which is defended by Deutsch (1997), Wallace (2012), and others; (2) Linde's (1986) *eternal inflation* view, which may have been confirmed by the results that Kuo related to Linde and Kallosh on their doorstep; (3) Smolin's (1997) *fecund universe theory*, which proposes that universes are generated through black holes; (4) the *cyclic model*, recently defended using string/M theory by Steinhardt and Turok (2007), which holds that distinct universes are formed in a never-ending sequence of Big Bangs and Big Crunches; and (5) Tegmark's (2007) 'Level IV' multiverse, which posits many universes governed by distinct mathematical and scientific laws. The details, implications, and overall scientific standing of these are theories are widely contested at present.[7]

This volume is not about the historical antecedents of these views, nor is it about the contemporary scientific debate concerning them. Instead, it is primarily about the role played by multiverse theories in certain current debates in philosophy of religion. Before introducing these, I will briefly mention—in order to then set aside—one philosophical context in which multiverses feature prominently. In contemporary analytic philosophy of religion, multiverse theories are frequently discussed in connection with *fine-tuning* arguments for the existence of God. There are many such arguments, but they all have the following basic structure. They begin by noting that if certain features of our universe had been slightly different, it would not have been capable of generating and sustaining life. They then claim that this apparent fine-tuning is best explained by the hypothesis of an intelligent designer who intentionally framed the universe to be biophilic. These arguments are, of course, hugely controversial. The most important criticism holds that they are undermined by multiverse theories. The basic idea is that if there are vastly many universes which vary (perhaps randomly) with respect to the relevant parameters, then it should not come as a surprise that at least one universe is life-permitting. So, in this context, multiverse theories are typically proffered as naturalistic *rivals* to theism.[8]

In recent years, however, several philosophers of religion have independently suggested that, far from being hostile to theism, multiple universes are what just what we should expect to find if God exists. This volume collects together twelve new essays that address this issue. The first two are by physicists, and the remaining ten are by philosophers. Before situating these essays with respect to the contemporary literature, and summarizing their arguments, I first set out some important aspects of the philosophical framework within which this discussion takes place. This will serve to set the stage.

THEISM, POSSIBLE WORLDS, ACTUALIZATION, AND CREATION

Much of contemporary analytic philosophy of religion addresses questions concerning the existence, nature, and activity of God. There are, of course, many models of God. Among philosophers, one of the most influential models holds that God is a necessarily existing being who cannot be surpassed in power, knowledge, and goodness, and who is the creator and sustainer of all that is.[9] Hereafter, I take *theism* to be to the claim that such a being exists.[10] In the remainder of this section, I introduce a way of thinking about the idea of God as the creator and sustainer.[11] This will provide a basis for the subsequent section's discussion of axiology.

Philosophers of religion often employ the language of *worlds* to illuminate the idea that God is the creator and sustainer of all that is. In this parlance, one can say (very roughly) that the *actual world* is everything that really exists, whereas each *possible world* is a unique way that things might have been. There are many accounts of what possible worlds are. Some philosophers think they are concrete objects (e.g., Lewis 1986), while others say they are abstract objects (e.g., Plantinga 1974), and still others deem them to be convenient fictions (e.g., Rosen 1990). The details and implications of these and other views about possible worlds are controversial, and there is no consensus about which one is correct.[12] On any of these views, however, one can say that if theism is true, God surveys the landscape of possible worlds, and then selects ours to be the actual world.[13] This process is called *actualizing* a world.

When considering the idea that God actualizes a world, it can be tempting to imagine (1) that God stands outside this landscape of possible worlds; that (2) God always *creates* something; and that (3) God determines each and every feature of the ensuing world (4) all at once; and that (5) God can choose any logically possible world for actualization. The first of these claims is false, and the rest are often denied by philosophers of religion.[14]

As noted, theism includes the claim that God is a necessary being: one who could not possibly fail to exist, or, equivalently, one who exists *in* all possible worlds. On this view, no sense can be made of the idea that God stands *outside* of the ensemble of worlds in order to select one for actualization. Since the possible worlds there are exhaust the way that things could be, there simply is no vantage point, divine or otherwise, entirely outside this ensemble.[15]

Second, while it is tempting to conflate world-actualization and *creation*, it is important to keep them distinct. Creation occurs, let's say, when God causes some spatiotemporal entity to be actual, but it's not the case that every instance of world-actualization involves this. Suppose that God exists, but creates nothing at all. If so, there still is an actual world. We might call it the *bare world,* since it is empty—except for whatever uncreated entities (such as God, and perhaps numbers) it contains. God, of course, is still responsible for the bare world's being actual, and so it makes sense to say that God has *actualized* it, without *creating* anything.

Third, God's actualizing a world need not mean that he determines each and every feature of the resulting world. Consider, for example, random processes. If a world includes such processes, then God causes it to be the case that they occur, but he does not (by definition) determine their outcome. Next, consider libertarian freedom.[16] Many theists maintain that human beings possess this kind of freedom, and that their free choices affect how the world unfolds. On this view, any world containing such creatures is *jointly* actualized by them and God. God is responsible, *inter alia,* for such a world's being the way it is prior to the introduction of creatures, and God is also responsible for the introduction of such creatures. But when they are introduced and begin to act freely, they too help make it the case that one world rather than another is actual. The resulting world, then, is partly the product of God's actions, and partly the product of creatures' actions.[17]

Fourth, there is no need to suppose that God's causal activity in actualizing a world is limited to one act at the (temporal or logical) beginning of that world. Some theists hold that God *intervenes* from time to time, and on this view, God performs many world-actualizing actions throughout the history of the world being actualized. In addition, as we have seen, theism holds that God's world-actualizing activity includes *sustaining* whatever is actual. This also suggests that this activity does not occur all at once at the outset of a world.

Finally, while it is tempting to suppose that God, given his omnipotence, can actualize any logically possible world, reasons have been offered for thinking otherwise. Consider, first, that there seem to be very bad logically possible worlds. One might think that while an omnipotent being would have the power to actualize such a world, a perfectly good being simply could not do so. On this view, such worlds, while logically possible, cannot be actualized by God.[18] Another influential reason for thinking that God cannot actualize every logically possible world is offered by Plantinga (1974). Plantinga claims that there are true propositions about how libertarian-free creatures will behave if placed in certain circumstances.[19] Although it is logically possible for these creatures to be in the relevant circumstances and to behave otherwise, Plantinga argues, not even God could actualize a world in which they find themselves in those circumstances, but freely act in ways other than those specified by these propositions. Given considerations like these, discussions of God's choice of a world often restrict their attention to those worlds that are *actualizable by God.*[20] In what follows, I presume this restriction.

WORLDS AND AXIOLOGICAL STATUS

So, philosophers of religion express the idea that God is the creator and sustainer by saying that God surveys the logical space of worlds within his power to actualize, and then selects exactly one world for actualization. But on what basis does God choose? Many philosophers have held that God chooses on the basis of the *objective value* of the worlds at issue. This section introduces a way of thinking and talking about the overall axiological status of worlds.

Consider such judgments as "it would have been far better had the Holocaust not happened", or "things would be far worse if slavery had not been abolished". These axiological evaluations aim to compare history as it really unfolded with one (or perhaps more) *counterfactual* histories: series of events that might have happened, but did not. At their broadest, such claims can be construed as comparative axiological judgments about worlds. Taken this way, to assert the former claim is to say that at least one possible world is better than the actual world, and to assert the latter is to say that at least one possible world is worse than the actual world. It is widely assumed in the relevant literature that worlds can coherently be supposed to have both absolute and relative axiological status.[21] If so, one can sensibly say that one world is good while another is bad, or that one world is better than or worse than another.

One way to spell out such claims is to say that a world, if actual, can bear, or fail to bear, *world-good-making* properties (hereafter WGMPs). These are properties that, *ceteris paribus,* tend to make worlds good.[22] Equally, one might say that a world, if actual, can bear, or fail to bear, world-*bad*-making properties (hereafter WBMPs).[23] If this is correct, then the overall axiological status of a world can be understood to depend upon which WGMPs and WBMPs are instantiated in the world, and, for degreed properties, the degree to which they are exemplified.[24]

This, of course, is just a framework for grounding absolute and comparative judgments of world-value. I have said nothing about which properties really are WGMPs or WBMPs, or about how they function, individually or jointly, or about what kinds of worlds really are good or bad. There are many different philosophical accounts of value, and these will have their own views about which properties are world-good-making or world-bad-making (and to what degree), and about whether and to what extent these properties can be jointly instantiated in a world, and about how they individually or jointly contribute to the overall axiological status of a world. A complete account would, presumably, settle these disputes. Moreover, such an account would also clarify the modal status of these properties, and would also reveal whether all worlds can really be compared, or whether, as some have held, there are genuine failures of comparability between worlds.[25] Of course, even if each world has an overall axiological status, and many or all pairs of worlds can be compared, this does not entail that finite creatures such as ourselves are always (or ever) capable of making the relevant judgments. But it is often taken for granted that God—an essentially omnipotent and omniscient being—would be able to make these judgments infallibly.[26] In the next section, I turn to some issues surrounding God's choice of a world.

GOD'S CHOICE OF A WORLD

So far, then, the overall picture looks like this. Theists say that God is the creator and sustainer of all that is. Analytic philosophers of religion express this idea by saying that God selects one world for actualization, and that he does so on the basis of its axiological properties. Suppose, for the moment, that all worlds can sensibly be compared with respect to value, that is, that there are no incomparable worlds. On this view, there are three distinct models of the hierarchy of possible worlds: either there is exactly one best of all possible worlds (as Leibniz famously thought), or else there are multiple unsurpassable worlds, or else there are no unsurpassable worlds, but instead an infinite hierarchy of increasingly better worlds. There are many thorny issues surrounding God's choice of a world on each of these hierarchies, and I briefly survey these now.[27]

Consider the first view, according to which there is one unique best of all possible worlds. It is natural to think that an omnipotent, perfectly good being will choose that world for actualization.[28] But if, given his attributes, God cannot fail to choose the best world, then one might wonder whether God's choice counts as free.[29] If it doesn't, one might wonder whether God's world-actualizing action really is worthy of thanks and praise, as theists typically suppose it to be. Moreover, if God cannot fail to choose the best world, one might wonder whether there really are any other possible worlds. Let's call these *the problem of divine freedom, the problem of thanks and praise,* and *the problem of modal collapse,* respectively. These can be thought of as in-house problems for theists. They can also be formulated as arguments for atheism.[30]

Next, consider the second view, according to which there are multiple—and perhaps even infinitely many—unsurpassable worlds. Variants of the three problems just noted seem to arise here as well: if, as seems natural to suppose, God cannot fail to choose an unsurpassable world, one might wonder whether God's choice counts as free, whether God's action is worthy of thanks and praise, and one might also wonder whether there really are any surpassable worlds at all. In addition, there are further questions to consider on this view: assuming that God will choose an unsurpassable world, which one will he choose, and how, and can his choice be deemed rational?[31] Again, these questions can be deemed to pose in-house problems for theists, but they can also motivate arguments for atheism.

Finally, consider the third view, according to which there is an infinite hierarchy of increasingly better possible worlds. Several authors have argued, in various ways, that this view precludes theism. The basic idea is this: theism maintains that God is essentially unsurpassable, but no matter which world from the hierarchy God chooses, God could have selected a better one, in which case God is surpassable in either rationality, or goodness, or both. And this, of course, is inconsistent with the idea that God is essentially unsurpassable in these respects. *A priori* arguments for atheism along these lines in this vein can be grouped under the heading *the problem of no best world.*[32]

Until now, we have supposed that all worlds can sensibly be compared with respect to value. If, however, this is false, further puzzles arise for the idea that God chooses one possible world to actualize. On what basis can God choose between incomparable alternatives? Various answers have been proposed, but it is sometimes argued that no choice between incomparable alternatives can be rational, and that this counts against theism if indeed there are genuine failures of comparability between worlds.[33]

There is one further issue concerning God's choice of a world that must be mentioned here. Whether there is one unsurpassable world, or multiple unsurpassable worlds, or no unsurpassable worlds, critics of theism can argue (and have argued) that the actual world is surpassable. When such arguments invoke *a posteriori* premises about the existence, nature, scope, duration, or distribution of some type or token of evil, they belong to a family of arguments for atheism collectively known as the *problem of evil.* An enormous literature in contemporary analytic philosophy of religion concerns such arguments,[34] perhaps because the problem of evil is thought to be the strongest objection to theism.[35] As we will see, theistic multiverse theories are typically deployed in responses to arguments from evil.

WORLDS AND UNIVERSES

It is important to distinguish *worlds* from *universes*. It can be useful to speak of smaller domains within a world, and to deem some of these to be *universes*.[36] To say that a multiverse is possible, then, is just to say that at least one possible world features two or more universes. Whether the *actual* world really includes multiple universes is, of course, a complex and vexed question. Different scientific theories offer different accounts of what constitutes a universe, and, as noted, there is considerable controversy within the scientific community about whether there really are multiple universes. Most philosophers who have proffered or discussed theistic multiverse theories take universes to be large, spatiotemporally interrelated objects that cannot interact with each other (e.g., Turner 2003; O'Connor 2008; Kraay 2010b). Some appear to assume that universes are related to each other, either temporally (Stewart 1993), or by being embedded in a higher spatial dimension (Hudson 2006). Others are officially neutral on these issues (Forrest 1996; Draper 2004).

As we earlier restricted our attention to possible worlds *actualizable* by God, let's now restrict our attention to universes *creatable* by God.[37]

Proponents of theistic multiverse theories typically believe that, like worlds, universes can sensibly be thought to have objective axiological status. Adapting the account given earlier, we might elaborate this idea with reference to a set of *universe*-good-making properties (UGMPs), and a set of *universe*-bad-making properties (UBMPs), such that the axiological status of universes depends upon which of these are instantiated, and, for degreed properties, to what degree.[38] Most theistic multiverse theories appear to assume that all universes can be compared with respect to value.[39] Some philosophers think it obvious that there are unsurpassable universes (e.g., McHarry 1978), while others hold that there are no unsurpassable universes (e.g., O'Connor 2008).

CONTEMPORARY THEISTIC MULTIVERSE THEORIES

The contemporary literature in analytic philosophy of religion on theistic multiverses begins with McHarry (1978), Forrest (1981), Parfit (1991, 1992), and Stewart (1993), all of whom briefly suggest that, in response to arguments for atheism that appeal to evil, theists could speculate that God has actualized a multiverse comprised of universes above some objective axiological threshold. The basic intuition these authors share is that God might reasonably be expected to create many such universes (assuming this is possible), and that it is more difficult to establish the surpassability of the multiverse as a whole than of one universe in particular. More recent and more developed proposals can be found in Forrest (1996), Turner (2003), Hudson (2006, 2013), Collins (2007), O'Connor (2008), Kraay (2010b, 2011b, 2012, 2013), Megill (2011), and Gellman (2012).[40]

Theistic multiverse proposals are controversial. McHarry (1978) is criticized by Perkins (1980) and Monton (2010). Forrest (1996) is criticized by Monton (2010). Turner (2003) is criticized by Almeida (2008, 2010), Monton (2010), Kraay (2012), and Pruss (forthcoming). Hudson (2006) is criticized by Almeida (2008), Rea (2008), and Monton (2010). O'Connor (2008) is criticized by Oppy (2008), Mawson (2009), Almeida (2010),[41] Craig (2010), Monton (2010), and Johnson (forthcoming). Kraay (2010b) is criticized by Monton (2010), Ijjas, Grössl, and Jaskolla (2013), Johnson (forthcoming), and Pruss (forthcoming). And Megill (2011) is criticized by Kraay (2013). Space does not permit a detailed discussion of every move and countermove in this complex debate.[42] Instead, in the remainder of this section, I will set out some of the key features of the theistic multiverses that have been proposed, and the uses to which they have been put, along with some of the most important objections they face.

Defenders of theistic multiverses all maintain that an essentially omnipotent, omniscient, and perfectly good God will create only those universes that surpass some objective axiological threshold. Most maintain that God will create *every* universe above this threshold.[43] O'Connor (2008, 119), in contrast, denies this,[44] and others are silent or neutral on this issue (Stewart

1993; Forrest 1981, 1996).[45] Some defenders of theistic multiverses maintain that universes come in different *kinds* or *types,* and that God will choose at least one of each (Stewart 1993; O'Connor 2008; Forrest 1981, 1996). Some authors claim that there can be *duplicate* universes within a multiverse (Parfit 1992; Monton 2010), but others deny this, citing the Principle of Identity of Indiscernibles (McHarry 1978; Turner 2003). Still others are silent or neutral on this issue (Stewart 1993; Forrest 1981, 1996; Hudson 2006; O'Connor 2008; Kraay 2010a).

Some authors think that a theistic multiverse comprising all and only those universes objectively worthy of being created and sustained by God is the unique best of all possible worlds (e.g., Turner 2003; Hudson 2006, 2013; Kraay 2010b). If this is plausible, then the problem of no best world cannot arise, nor can the problem of how God is to choose between multiple unsurpassable worlds.

But arguments for the claim that such a multiverse is the unique best possible world face two major challenges. The first concerns the threshold of universe worthiness. Various construals of this threshold have been developed, and, predictably, these are controversial.[46] Moreover, it has been argued that if there are no unsurpassable universes, then it is incoherent to suppose that *any* threshold could be acceptable for an unsurpassable being to choose, since for any threshold one might specify, a higher one could be defended (Johnson, forthcoming).

The second major challenge holds that a multiverse of *all* worthy universes is logically impossible. Several philosophers appeal to considerations about personal identity in support of this claim.[47] Others argue that for any number of universes God could create, God could create even more—and that, accordingly, no multiverse can house them all (Monton 2010, Johnson forthcoming). It has also been argued that there are pairs of worthy universes such that only one can be included.[48] Finally, it has been argued that the idea of such a multiverse conflicts with the divine omniscience (Pruss, forthcoming), thus generating a contradiction for theistic multiverses.[49]

Most defenders of theistic multiverses believe that their model will help theists respond to arguments from evil. As noted earlier, this is the primary application of these theories. They differ, though, in their estimation of how significant this assistance will be. Some think that multiverses can make a modest contribution to theistic responses (e.g., O'Connor 2008). Others think it has the resources to significantly enhance existing theistic strategies (e.g., Hudson 2013).[50] The limit case is Megill (2011), who believes that the bare epistemic possibility of multiple universes completely defeats all arguments from evil, past and present. All these claims are controversial. Draper (2004), Almeida (2008), and Monton (2010) have argued, in various ways, that theistic multiverse theories cannot defeat arguments from evil, and I have argued likewise (Kraay 2012, 2013).[51]

Before turning to the present volume, a final word is needed about theistic multiverses and puzzles for theism. Even if it can be shown that some model of a theistic multiverse is logically possible, and indeed that it is the best of all possible worlds, and even if it is reasonable to believe that the actual world is (or probably is) such a multiverse, so that the problem of evil is either mitigated or resolved entirely, some of the puzzles pertaining to God's choice of a world would still remain. Theists would still have to address three problems noted earlier: the *problem of divine freedom,* the *problem of thanks and praise,* and the *problem of modal collapse.*[52]

CHAPTER SUMMARIES

This volume opens with two chapters by physicists. In "Puzzled by Particularity", Robert B. Mann argues that multiverse hypotheses spell trouble for both science and theism. After setting out some surprising features of our universe, including its biophilic character, he distinguishes four ways of accounting for them: randomness, cosmic necessity, intelligent design, and, finally, the postulation that our universe is part of a much larger multiverse. Mann concedes that there is indirect scientific support for the final view, but he thinks its attractiveness is merely superficial. He argues that multiverse theories involve two features (rampant duplication and Boltzmann brains) that severely compromise scientific inquiry. He also claims that multiverse theories are difficult to square with theism, in part because they are committed to the actual existence of massive quantities of evil, all of which are repeated arbitrarily many times.

In contrast to Mann, Don N. Page takes a much brighter view of both the scientific and theological potential for at least one multiverse model: the multiple 'worlds' postulated by the Everett interpretation of quantum theory. Page begins "The Everett Multiverse and God" by setting out some simplicity-based considerations that, he thinks, favor this interpretation. Page then entertains what he takes to be an even simpler explanation of why the world is as it is: the hypothesis that the actual world is the best possible world. Page supposes that the best possible world would be one that maximizes the intrinsic value of conscious experience, and he speculates that such a world would have to include an omnipotent, omniscient, omnibenevolent creator who has enormous appreciation for the mathematical elegance of the universe.[53] Page then argues that it is plausible to suppose that such a being would bring about an Everett multiverse.

The remaining contributions to this volume are all by philosophers. The essays in the next section explore various details and applications of theistic multiverses. Peter Forrest's chapter is titled "The Multiverse: Separate Worlds, Branching, or Hyperspace? And What Implications Are There for Theism?" Forrest thinks that multiverse views can offer good accounts of agency, freedom, time, and probability. After distinguishing several distinct multiverse theories, Forrest says that theists should prefer a view called *hyperspace,* on which universes are four-dimensional subspaces of a larger overall structure containing more—and perhaps many more—dimensions. He closes by expressing tentative support for a version of pantheism, according to which God just is the actual universe. (Later chapters by Nagasawa and Leslie consider different forms of pantheism.)

Jason L. Megill's chapter, "An Argument for Modal Realism", defends the claim that there are at least two worlds containing literally concrete entities. This view is inconsistent with the view that all possible worlds are abstract objects, but is consistent with (although considerably more modest than) the modal realism of Lewis (1986), which holds that there are infinitely many concrete worlds. Megill's argument contains only two premises. The first is "if an entity *e* is possibly literally concrete in the actual world, then there is a possible world *w* in which it is literally concrete", and the second is "there (1) is an entity *e* that is possibly literally concrete in the actual world but (2) *e* is not literally concrete in the actual world". Megill offers three arguments in favor of each premise, and then considers what bearing his view has for theism. He argues that theists should favor a restricted modal realism according to which (1) God ensures that there are no universes unworthy of divine creation, and (2) there are not so many universes that fine-tuning arguments for theism are undermined.

The final essay in this section is by Donald A. Turner. In an important previous publication, Turner (2003) argued that the hypothesis of many universes can be deemed a partial solution to the problem of evil. In this chapter, "Revisiting the Many-Universes Solution to the Problem of Evil", Turner responds to several objections to his view. Contra Monton (2010), Turner argues that an omnipotent, omniscient, perfectly good God would create every universe with a favorable balance of good over evil—and if duplicates are possible, God would create all possible duplicates of these as well. Turner next addresses a series of objections due to Almeida (2008), several of which claim that Turner's multiverse is logically impossible. He resists Almeida's claim that his multiverse is the only possible world, but concedes to Almeida that his view entails that God is not free to select any other world, and that no evil is genuinely gratuitous. Finally, Turner disagrees with my judgment (expressed in Kraay 2012) that his multiverse theory will not significantly help theists respond to the problem of evil.

The next section of the volume includes three essays that are (largely) critical of multiverse theories. In a previous publication, I argued that theists should expect the actual world to be a theistic multiverse (TM) comprising all and only those universes that are worthy of being created and sustained by God (Kraay 2010b). I further claimed that this would be the unique best possible world. Accordingly, I urged, this view would evade two *a priori* arguments for atheism that depend upon there being no such thing. I have also argued that this view will simply reframe existing debates concerning *a posteriori* arguments for atheism, without advantaging either side (Kraay 2010b, 2012). Michael Schrynemakers's chapter, "Kraay's Theistic Multiverse", responds to these claims. He argues that no divinely furnished multiverse can possibly include *all* worthy universes, in which case there is no reason to suppose that TM is the unique best of all possible worlds, and, accordingly, no way for theists to appeal to TM to evade these *a priori* arguments for theism. Schrynemakers also argues, however, that a model like TM will make it more difficult to argue for atheism *a posteriori* by appealing to the surpassability of the actual world in general, or to the presence of evil in it in particular. In this respect his position is similar to that of Turner.

Michael Almeida agrees with Michael Schrynemakers that a multiverse comprising all and only worthy universes is not logically possible, but he offers a different argument for this conclusion. In his chapter, "Best Worlds and Multiverses", Almeida assumes for *reductio* that necessarily God actualizes the best possible world—a world containing, inter alia, *all* universes in which all moral agents always observe all requirements of justice and beneficence and *no* universes in which all moral agents always violate all requirements of justice and beneficence. If, necessarily, God actualizes this world, then there are no alternate possibilities. In particular, it is not metaphysically possible for there to be universes in which all moral agents always violate all requirements of justice and beneficence. But, Almeida argues, such universes must be metaphysically possible in order for there to be universes in which all moral agents always observe all requirements of justice and beneficence. This is because observing all requirements of justice and beneficence requires significant moral freedom, and this sort of freedom, Almeida thinks, requires the existence of alternative metaphysical possibilities. Accordingly, he thinks, there can be no theistic multiverse comprising all and only worthy universes.

Jeremy Gwiazda's chapter, "On Multiverses and Infinite Numbers", considers this question: "How many universes are there in the multiverse?" Gwiazda argues that a nonstandard conception of infinite numbers, on which infinite numbers behave very much like finite numbers, is preferable to the Cantorian view of infinite numbers. He then argues that this conception lowers the prior probability of there being a multiverse comprising infinitely many universes. This, of course, is a far more modest criticism of theistic multiverse theories than the ones leveled by Schrynemakers and Almeida.

The next section of the volume explores pantheistic views of ultimate reality. In "Multiverse Pantheism", Yujin Nagasawa distinguishes traditional pantheism (the view that God is identical with our universe) from multiverse pantheism (the view that God is identical with the multiverse posited by Lewisian modal realism). He then sets out three objections to the former view: the universe cannot be God because (1) it is finite, while God is supposed to be infinite; (2) it contains evil, but God is supposed to be perfect; and (3) it is unworthy of worship, whereas God is supposed to be worshipworthy. Nagasawa then argues that multiverse pantheism fares better than does traditional pantheism against objections (1) and (3), but that it is vulnerable to a version of (2).

In his chapter, "God and Many Universes", John Leslie defends a world-view that one might call pantheistic idealism.[54] Whereas the pantheism Nagasawa considers treats our universe as a concrete object, on Leslie's view, what we call our universe just is a thought pattern contemplated by a divine mind. Moreover, there are infinitely many other universes, each of which just is a thought pattern contemplated by another divine mind. Leslie thinks that these minds exist because it is ethically or axiologically good for them to exist. Leslie is neutral between four ways of using the term 'God' on this worldview: God might be (1) the entire multiverse of divine minds, taken together; (2) the mind whose thoughts compose *our* universe; (3) the principle that the ethical/axiological need for existence is creatively power-ful; or (4) an all-seeing, personality-imbued region of an infinite mind. After setting out some further details of his worldview, Leslie argues that it can be defended against the problem of evil.

The final two essays consider what bearing multiverses might have for the Christian doctrine of the Incarnation.[55] In his chapter, "Extraterres-trial Intelligences and the Incarnation", Robin Collins first offers reasons for supposing that there are many other nonhuman 'races' of vulnerable, embodied conscious agents (VECAs). He then argues that Christians have good reason to believe that many of these have, like ourselves, fallen, and that it is extremely improbable that God would become incarnate only in our own race. Thus Christians should be motivated to develop a theologi-cally satisfying model of multiple incarnations. He then considers several models of the Incarnation, and argues that all but one of them are entirely compatible with multiple incarnations, and that even this one can be made so, given certain assumptions concerning time. In their chapter, "Incarna-tion and the Multiverse", Timothy O'Connor and Philip Woodward sug-gest similar motivations for developing a model of multiple incarnations. They set out and defend a version of a compositional theory, according to which an incarnate deity has two natures, each of which is a distinct component of its being. They then extend this model to permit multiple incarnations. Finally, they consider an objection to this model based on the theological idea that Christ's work is necessary for ushering in a united community of all divine-image-bearing creatures. In response, they specu-late that no such all-encompassing community would be possible, given the vast differences between such creatures. Accordingly, they speculate that each incarnation could help to bring about a unified community of the relevant sort of divine-image-bearing creatures, and that each of these com-munities would, in its own way, participate in the common goal of union with God.

ACKNOWLEDGMENTS

Thanks are due to Ryerson University and its Department of Philosophy for supporting my research in many ways since I arrived in fall 2003. Thanks are due to the Department of Philosophy at Monash University, and to Graham Oppy in particular, for arranging office space for me and supporting my research from January to June 2011. Thanks are due to St. Peter's College, Oxford, and to Tim Mawson in particular, for arranging office space for me and supporting my research from September 2011 to June 2012. Thanks are due to the John Templeton Foundation for funding my Visiting Fellowship at Oxford and for funding the research workshop held at Ryerson University on February 15–16, 2013, at which ancestors of many essays in this volume were first presented. (I am especially thankful to Linda Zagzebski and Brian Leftow for ably managing the overall research project of which my fellowship and this research workshop formed small parts.) Thanks are also due to the contributors to this volume, not only for producing their fine essays, but also for generously giving valuable feedback on the work of other contributors. Finally, and most importantly, I am profoundly grateful to my wife, Mary Beth, for her unconditional love and for her unwavering support of this project during a hectic—but wonderful and unforgettable!—period of time during which we lived on three different continents and began to raise our two beautiful children.

NOTES

1. The video can be found at www.youtube.com/watch?v=ZlfIVEy_YOA. At the time of this anthology's publication, it had been viewed almost three million times.
2. More information can be found on the project website: www.cfa.harvard.edu/ CMB/bicep2/science.html. The as-yet-unpublished results were also released on March 17, 2014, and can be found at http://bicepkeck.org/. Press coverage of these results can be found in Amos (2014), Clark (2014), Grossman (2014), and Overbye (2014).
3. For more, see NASA's news release at www.jpl.nasa.gov/news/news.php? release=2014–082.
4. As quoted in Grossman (2014).
5. Amos (2014), Clark (2014), and Grossman (2014) report that this sentiment has been expressed by Alan Guth (MIT), Andrew Jaffe (Imperial College), and Avi Loeb (Harvard), respectively.
6. What follows is a mere sketch of this history. More comprehensive presentations can be found in Lovejoy (1936), Munitz (1951), Dick (1982), and Crowe (1996, 2008). I should stress that, down through the ages, thinkers have meant very different things by terms like 'world', 'cosmos', and 'universe' (and by the words variously translated into these), and this, of course, significantly complicates the relevant intellectual history.
7. For opinionated introductions of these and other multiverse theories, see Leslie (1989), Rees (2001), Kaku (2005), Susskind (2005), Vilenkin (2006), Carr (2007), Gribbin (2009), Greene (2011), Barrow (2012), and Wallace (2012).
8. Good introductions to these arguments, and to multiverse-based criticisms of them, can be found in Himma (2006), Collins (1999, 2006, 2007, 2009), Manson (2003, 2009), and Ratzsch (2010).
9. Good introductions to the wide diversity of models of God currently discussed by analytic philosophers include Diller and Kasher (2013) and Nagasawa and Buckareff (forthcoming).

10. Most papers in this volume engage with this view of God, but those by Nagasawa and Leslie (Chapters 9 and 10, respectively) instead consider a *pantheistic* conception of God.
11. For more on this topic, see Kraay (2008) and Kraay, Chantler, and Lougheed (forthcoming).
12. Good entry points into the vast literature on possible worlds are Divers (2002), Shalkowski (2011), Parent (2012), and Menzel (2013).
13. There is some controversy in the literature concerning whether theism is compatible with the modal realism of Lewis (1986). For a survey of this issue, see the section entitled "God and Modal Realism" in Kraay, Chantler, and Lougheed (forthcoming).
14. The next four paragraphs are adapted from Kraay (2008) and Kraay (2010b).
15. In Chapter 5 in this volume, Turner says that God is outside of all worlds, but he means something different by 'world': "a single maximal spatiotemporal aggregate, a cosmos or universe" (114). On his view, there is a vantage point outside the set of all these.
16. A good entry point into the massive literature concerning free will is Kane (2011).
17. On this point, see Plantinga (1972, 169–90).
18. On this issue, see Guleserian (1983) and the other papers cited under the heading "God and Bad Worlds (The Modal Problem of Evil)" in Kraay, Chantler, and Kougheed (forthcoming).
19. These have come to be called *counterfactuals of creaturely freedom*. For good introductions to the debate about such claims, see Flint (2009) and Zagzebski (2011). For some of the latest moves, see the papers collected in Perszyk (2011).
20. An alternative move here (and one that I favor) is to claim that these putatively-possible-but-not-divinely-actualizable worlds are not, after all, genuine possibilities, given theism. Thomas Morris (1987, 48), for example, says that

> God is a delimiter of possibilities. If there is a being who exists necessarily, and is necessarily omnipotent, omniscient, and good, then many states of affairs which otherwise would represent genuine possibilities, and which by all non-theistic tests of logic and semantics do represent possibilities, are strictly impossible in the strongest sense. In particular, worlds containing certain sorts of disvalue or evil are metaphysically ruled out by the nature of God, divinely precluded from the realm of real possibility.

21. One explicit discussion of this issue is Menssen (1996).
22. Candidate WGMPs pick out a property held to be good-making. Traditional examples include the presence of free moral agents in the world, the favorable balance of moral actions over immoral ones, the variety of phenomena in the world, and the simplicity of a world's governing laws.
23. Candidate WBMPs typically appeal to the presence of unjustified evil or suffering in the world. On the Augustinian view, according to which evil is in fact the absence of good (*privatio boni*), every WBMP would presumably refer to the *absence* of a WGMP. There may be such WBMPs, and there may also be WBMPs that are the *contraries* of WGMPs, and there may be other, different, WBMPs.
24. It may be that certain good-making properties cease to make worlds better past a certain point, or in certain combinations. The same goes, *mutatis*

mutandis, for WBMPs. So, while the goodness of a world depends on its axiological properties, this dependency may not be simple.

25. I discuss both issues in Kraay (2011a, 2012). For a survey of literature relevant to the latter issue, see Kraay, Chantler, and Lougheed (forthcoming).

26. One important dissenting view is called *Open Theism*. For discussions that situate this view relative to alternatives, see Flint (2009) and Zagzebski (2009). For an influential defense of it, see Pinnock et al. (1994).

27. For discussions of the key literature on these three issues, see Kraay (2008) and Kraay, Chantler, and Lougheed (forthcoming).

28. Adams (1972) denies this, but his argument has been widely criticized. One particularly clear critic is Rowe (2004).

29. On divine freedom, see Rowe (2004) and Timpe (2014).

30. Such arguments proceed by claiming, first, that theists cannot plausibly give up (some relevant conception of) divine freedom, or divine thankworthiness and praiseworthiness, or the claim that there really are sub-optimal possibilities. They then urge that one or more of these is incompatible with the idea that God cannot fail to choose the unique best world.

31. For further discussion, see Kraay (2008).

32. For further discussion, see Kraay (2010a).

33. For further discussion, see Kraay (2011a).

34. A good starting point is McBrayer and Howard-Snyder (2013).

35. For results of a 2012 survey showing this, see de Cruz (2014).

36. Some contemporary authors, however, use 'universe' and 'world' in precisely the opposite way, and others use different terminology altogether. For consistency and simplicity, I will employ the dominant nomenclature.

37. It is generally thought that possible worlds (unlike universes) can neither be created nor destroyed, which is why the term 'actualize' is used for God's activity in making a world actual. For clarity, then, it is useful to reserve the term 'create' for universes.

38. I suggested earlier that the dependency of the axiological status of a world on its WGMPs and WBMPs need not be simple. Similarly, the dependency of the axiological status of *universes* on the relevant properties need not be simple. One further point. Some WGMPs can equally be deemed *universe-good-making* properties. But not all: consider the property *comprising many good universes*. While this is a plausible *world*-good-making property, it cannot be a *universe*-good-making property.

39. One exception is O'Connor (2008).

40. One physicist, Don Page, has also expressed his sympathy for the idea of a theistic multiverse (Page 2010; see also Chapter 2 in this volume).

41. O'Connor (2010) responds.

42. For an annotated guide to this literature, see Kraay, Chantler, and Lougheed (forthcoming). For an opinionated survey of it, see Kraay (2012).

43. See McHarry (1978), Turner (2003), Hudson (2006, 2013), and Kraay (2010b).

44. O'Connor's denial that God will create every universe above the threshold has been criticized in various ways (Almeida 2010; Monton 2010).

45. See also Draper (2004).

46. I survey this discussion in Kraay (2012).

47. As recorded by Turner (2003), Pruss advances this objection. As recorded by Kraay (2012), Peter van Inwagen and Tom Talbott also offered it. It is also advanced by Almeida (2008, Chapter 7 in this volume) and discussed in Lougheed (forthcoming).

48. I discuss one such objection due to Pruss in Kraay (2012).

49. The essays by Schrynemakers and Almeida (Chapters 6 and 7 in this volume) offer new arguments for the logical impossibility of a theistic multiverse comprising all and only universes worthy of being created and sustained.
50. Schrynemakers's essay (Chapter 6 in this volume) also argues for this claim.
51. To date, attempts to deploy theistic multiverse theories in response to arguments from evil have generally not engaged with the latest developments in the debate concerning the latter. Perhaps this is why Hudson (2013, 246) deems this work "promising but underdeveloped".
52. These arguments, of course, can be leveled against any view of what this best world is—they do not just target multiverse models.
53. Page here inverts the traditional argument, made by Leibniz and others, which holds that *since* God exists, the actual world must be the best of all possible worlds. A similar appeal to multiverse-based aesthetic considerations can be found in Hudson (2006).
54. Leslie is the leading contemporary proponent of this view, having defended it at length elsewhere, notably in Leslie (2001).
55. This is not, I should stress, an entirely new question for Christian theology. Shortly after Copernicus published *De Revolutionibus Orbium Coelestium* in 1543, the Lutheran reformer Philipp Melanchthon challenged heliocentrism by warning that it might foster either the unorthodox idea that Christ's redemption would be unnecessary for denizens of other worlds, or else the absurd idea that Christ would manifest himself in such worlds. Many other thinkers have also grappled with this issue. (For more, see Dick 1982; Crowe 1996, 2008). In the contemporary literature on multiverses in analytic philosophy of religion, however, only Hudson (2006, Chapter 8) considers how one multiverse view (his theory of hyperspace) bears on Christian doctrine.

REFERENCES

Adams, R. M. 1972. "Must God Create the Best?" *The Philosophical Review* 81: 317–32.
Almeida, M. 2008. *The Metaphysics of Perfect Beings.* New York: Routledge.
Almeida, M. 2010. "O'Connor's Permissive Universe." *Philosophia Christi* 12: 296–307.
Amos, J. 2014. "Cosmic Inflation: 'Spectacular' Discovery Hailed." *BBC News— Science and Environment,* March 17. www.bbc.com/news/science-environment-26605974.
Barrow, J. D. 2012. *The Book of Universes.* London: Vintage.
Carr, B., ed. 2007. *Universe or Multiverse?* Cambridge: Cambridge University Press.
Clark, S. 2014. "Gravitational Waves: Have US Scientists Heard Echoes of the Big Bang?" *The Guardian,* March 17. www.theguardian.com/science/2014/mar/14/gravitational-waves-big-bang-universe-bicep.
Collins, R. 1999. "A Scientific Argument for the Existence of God: The Fine-Tuning Design Argument." In *Reason for the Hope Within,* edited by M. Murray, 47–75. Grand Rapids, MI: Eerdmans.
Collins, R. 2006. "The Many-Worlds Hypothesis As an Explanation of Cosmic Fine-Tuning: An Alternative to Design?" *Faith and Philosophy* 22: 654–66.
Collins, R. 2007. "The Multiverse Hypothesis: A Theistic Perspective." In *Universe or Multiverse?,* edited by B. Carr, 459–80. New York: Cambridge University Press.
Collins, R. 2009. "The Teleological Argument: An Exploration of the Fine-Tuning of the Universe." In *The Blackwell Companion to Natural Theology,* edited by W. L. Craig and J. P. Moreland, 202–81. Malden, MA: Wiley-Blackwell.

Craig, W. L. 2010. "Timothy O'Connor on Contingency: A Review Essay on *Theism and Ultimate Explanation.*" *Philosophia Christi* 12: 181–88.

Crowe, M. J. 1996. *The Extraterrestrial Life Debate, 1750–1900.* Cambridge: Cambridge University Press.

Crowe, M. J. 2008. *The Extraterrestrial Life Debate, Antiquity to 1915: A Source Book.* Notre Dame, IN: University of Notre Dame Press.

Deutsch, D. 1997. *The Fabric of Reality: The Science of Parallel Universes and Its Implications.* Harmondsworth, UK: Penguin.

de Cruz, H . 2014. "Preliminary Results of the Survey on Natural Theological Arguments." www.academia.edu/1438058/Results_of_my_survey_on_natural_theological_arguments.

Dick, S. 1982. *Plurality of Worlds: The Origins of the Extraterrestrial Life Debate from Democritus to Kant.* Cambridge: Cambridge University Press.

Diller, J., and A. Kasher, eds. 2013. *Models of God and Alternative Ultimate Realities.* Dordrecht: Springer.

Divers, J. 2002. *Possible Worlds.* London: Routledge.

Draper, P. 2004. "Cosmic Fine-Tuning and Terrestrial Suffering: Parallel Problems for Naturalism and Theism." *American Philosophical Quarterly* 41: 311–21.

Everett, H. 1957. "'Relative State' Formulations of Quantum Mechanics." *Reviews of Modern Physics* 29: 454–62.

Flint, T. 2009. "Divine Providence." In *The Oxford Handbook of Philosophical Theology,* edited by T. Flint and M. Rea, 262–85. Oxford: Oxford University Press.

Forrest, P. 1981. "The Problem of Evil: Two Neglected Defences." *Sophia* 20: 49–54.

Forrest, P. 1996. *God without the Supernatural.* Ithaca, NY: Cornell University Press.

Gellman, J. 2012. "A Theistic, Universe-Based, Theodicy of Human Suffering and Immoral Behaviour." *European Journal for Philosophy of Religion* 4: 107–22.

Greene, B. 2011. *The Hidden Reality: Parallel Universes and the Deep Laws of the Cosmos.* New York: Knopf.

Gribbin, J. 2009. *In Search of the Multiverse.* London: Penguin.

Grossman, L. 2014. "Multiverse Gets Real with Glimpse of Big-Bang Ripples." *New Scientist,* March 18. www.newscientist.com/article/dn25249-multiverse-gets-real-with-glimpse-of-big-bang-ripples.html?page=1#.U2vNLfldVex.

Guleserian, T. 1983. "God and Possible Worlds: The Modal Problem of Evil." *Noûs* 17: 221–38.

Himma, K. E. 2006. "Design Arguments for the Existence of God." In *Internet Encyclopedia of Philosophy,* edited by J. Feiser and B. Dowden. www.iep.utm.edu/d/design.htm.

Hudson, H. 2006. *The Metaphysics of Hyperspace.* Oxford: Oxford University Press.

Hudson, H. 2013. "Best Possible World Theodicy." In *The Blackwell Companion to the Problem of Evil,* edited by J. McBrayer and D. Howard-Snyder, 236–50. Oxford: Wiley-Blackwell.

Ijjas, A., J. Grössl, and L. Jaskolla. 2013. "Theistic Multiverse and Slippery Slopes: A Response to Klaas Kraay." *Theology and Science* 11: 62–76.

Johnson, D. K. Forthcoming. "The Failure of the Multiverse Hypothesis as a Solution to the Problem of No Best World." *Sophia.*

Kaku, M. 2005. *Parallel Worlds: A Journey through Creation, Higher Dimensions, and the Future of the Cosmos.* New York: Random House.

Kane, R., ed. 2011. *The Oxford Handbook of Free Will.* Oxford: Oxford University Press.

Kraay, K. 2008. "Can God Choose a World at Random?" In *New Waves in Philosophy of Religion,* edited by E. Wielenberg and Y. Nagasawa, 22–35. Hampshire, UK: Palgrave Macmillan.

Kraay, K. 2010a. "The Problem of No Best World." In *A Companion to Philosophy of Religion*, 2nd ed., edited by C. Taliaferro, P. Draper, and P. Quinn, 481–91. Oxford: Blackwell.

Kraay, K. 2010b. "Theism, Possible Worlds, and the Multiverse." *Philosophical Studies* 147: 355–68.

Kraay, K. 2011a. "Incommensurability, Incomparability, and God's Choice of a World." *International Journal for Philosophy of Religion* 69: 91–102.

Kraay, K. 2011b. "Theism and Modal Collapse." *American Philosophical Quarterly* 48: 361–72.

Kraay, K. 2012. "The Theistic Multiverse: Problems and Prospects." In *Scientific Approaches to the Philosophy of Religion*, edited by Y. Nagasawa, 142–62. Houndsmills, UK: Palgrave Macmillan.

Kraay, K. 2013. "Megill's Multiverse Meta-Argument." *International Journal for Philosophy of Religion* 73: 235–41.

Kraay, K., A. Chantler, and K. Lougheed. Forthcoming. "God and Possible Worlds." In *Oxford Bibliographies Online*.

Leslie, J. 1989. *Universes*. New York: Routledge.

Leslie, J. 2001. *Infinite Minds: A Philosophical Cosmology*. Oxford: Clarendon Press.

Lewis, D. 1986. *On the Plurality of Worlds*. Oxford: Blackwell.

Linde, A. D. 1986. "Eternally Existing Self-Reproducing Chaotic Inflationary Universe." *Physics Letters, Series B* 175: 395–400.

Lougheed, K. Forthcoming. "Divine Creation, Modal Collapse, and the Theistic Multiverse." *Sophia*.

Lovejoy, A. O. 1936. *The Great Chain of Being: A Study of the History of an Idea*. Cambridge, MA: Harvard University Press.

Manson, N., ed. 2003. *God and Design: The Teleological Argument and Modern Science*. New York: Routledge.

Manson, N. 2009. "The Fine-Tuning Argument." *Philosophy Compass* 4: 271–86.

Mawson, T. 2009. Book Review of Timothy O'Connor's Theism and Ultimate Explanation. *Religious Studies* 45: 237–241.

McBrayer, J. and Howard-Snyder, D. 2013. *The Blackwell Companion to the Problem of Evil*. Oxford: Wiley-Blackwell.

McHarry, J. D. 1978. "A Theodicy." *Analysis* 38: 132–34.

Megill, J. 2011. "Evil and the Many Universes Response." *International Journal for Philosophy of Religion* 70: 127–38.

Menssen, S. 1996. "Grading Worlds." *Proceedings of the American Catholic Philosophical Association* 70: 149–61.

Menzel, C. 2013. "Possible Worlds." In *Stanford Encyclopedia of Philosophy*, edited by E. N. Zalta. http://plato.stanford.edu/entries/possible-worlds/.

Monton, B. 2010. "Against Multiverse Theodicies." *Philo* 13: 113–35.

Morris, Thomas. 1987. "The Necessity of God's Goodness." In *Anselmian Explorations: Essays in Philosophical Theology*, 42–69. Notre Dame, IN: University of Notre Dame Press.

Munitz, M. K. 1951. "One Universe or Many?" *Journal of the History of Ideas* 12: 231–55.

Nagasawa, Y., and A. Buckareff, eds. Forthcoming. *Alternative Models of God*. Oxford: Oxford University Press.

O'Connor, T. 2008. *Theism and Ultimate Explanation: The Necessary Shape of Contingency*. Melbourne: Wiley-Blackwell.

O'Connor, T. 2010. "Is God's Necessity Necessary? Replies to Senor, Oppy, McCann, and Almeida." *Philosophia Christi* 12: 309–316.

Oppy, G. 2008. "Book Review of Timothy O'Connor's *Theism and Ultimate Explanation*." *Notre Dame Philosophical Reviews*. http://ndpr.nd.edu/review.cfm?id=13406.

Overbye, D. 2014. "Space Ripples Reveal Big Bang's Smoking Gun." *New York Times,* March 17. www.nytimes.com/2014/03/18/science/space/detection-of-waves-in-space-buttresses-landmark-theory-of-big-bang.html.

Page, D. 2010. "Does God So Love the Multiverse?" In *Science and Religion in Dialogue,* vol. 1, edited by M. Y. Stewart, 380–95. Oxford: Wiley-Blackwell.

Parent, T. 2012. "Modal Metaphysics." In *Internet Encyclopedia of Philosophy,* edited by J. Feiser and B. Dowden. www.iep.utm.edu/mod-meta/.

Parfit, D. 1991. "Why Does the Universe Exist?" *Harvard Review of Philosophy,* Spring, 4–5.

Parfit, D. 1992. "The Puzzle of Reality: Why Does the Universe Exist?" *Times Literary Supplement,* July 3. (Reprinted in P. van Inwagen and D. Zimmerman, eds., 1998, *Metaphysics: The Big Questions,* 418–27, Oxford: Blackwell.)

Perkins, R. K. 1980. "McHarry's Theodicy: A Reply." *Analysis* 40: 168–71.

Perszyk, K., ed. 2011. *Molinism: The Contemporary Debate.* Oxford: Oxford University Press.

Pinnock, C., R. Rice, J. Sanders, W. Hasker, and D. Basinger. 1994. *The Openness of God: A Biblical Challenge to the Traditional Understanding of God.* Downers Grove, IL: InterVarsity Press.

Plantinga, A. 1972. *The Nature of Necessity.* Oxford: Clarendon Press.

Pruss, A. Forthcoming. "Divine Creative Freedom." In *Oxford Studies in Philosophy of Religion.* Oxford: Oxford University Press.

Ratsch, D. 2010. "Teleological Arguments for God's Existence." In *Stanford Encyclopedia of Philosophy,* edited by E. N. Zalta. plato.stanford.edu/entries/teleological-arguments/.

Rea, M. 2008. "Hyperspace and the Best World Problem." *Philosophy and Phenomenological Research* 76: 444–51.

Rees, M. 2001. *Our Cosmic Habitat.* Princeton, NJ: Princeton University Press.

Rosen, G. 1990. "Modal Fictionalism." *Mind* 99: 327–54.

Rowe, W. 2004. *Can God Be Free?* Oxford: Oxford University Press.

Shalkowski, S. 2011. "Modality." In *Oxford Bibliographies Online,* edited by D. Pritchard. www.oxfordbibliographies.com/view/document/obo-9780195396577/obo-9780195396577-0077.xml.

Smolin, L. 1997. *The Life of the Cosmos.* Oxford: Oxford University Press.

Steinhardt, P., and N. Turok. 2007. *Endless Universe: Beyond the Big Bang.* New York: Doubleday.

Stewart, M. Y. 1993. *The Greater Good Defence: An Essay on the Rationality of Faith.* New York: St. Martin's Press.

Susskind, L. 2005. *The Cosmic Landscape: String Theory and the Illusion of Intelligent Design.* New York: Little, Brown.

Tegmark, M. 2007. "The Multiverse Hierarchy." In *Universe or Multiverse?,* edited by B. Carr, 99–125. Cambridge: Cambridge University Press.

Timpe, K. 2014. *Free Will in Philosophical Theology.* New York: Bloomsbury.

Turner, D. 2003. "The Many-Universes Solution to the Problem of Evil." In *the Existence of God,* edited by R. Gale and A. Pruss, 1–17. Aldershot, UK: Ashgate.

Vilenkin, A. 2006. *Many Worlds in One: The Search for Other Universes.* New York: Hill and Wang.

Wallace, D. F. 2012. *The Emergent Multiverse: Quantum Theory According to the Everett Interpretation.* Oxford: Oxford University Press.

Zagzebski, L. 2011. "Foreknowledge and Free Will." In *Stanford Encyclopedia of Philosophy,* edited by E. N. Zalta. http://plato.stanford.edu/entries/free-will-foreknowledge/.

Physicists on God and
the Multiverse

1 Puzzled by Particularity

Robert B. Mann

It is now clear that our cosmos is riddled with a considerable degree of particularity. Responding to this, a number of scientists have in recent years advocated a "super-Copernican" revolution, in which our universe is regarded as a small part of a much larger structure known as the multiverse. Scientifically, this entails an unprecedented combination of broadened theoretical perspective with severe empirical limitations, implicitly redefining what is meant by science. Theologically, it introduces a new question: why is there something instead of everything? I give an overview of the epistemic costs the multiverse extracts for both science and theology.

GEOCENTRIC PRELUDE

If there is a singular lesson to be learned from the Copernican revolution, it is that we must be highly wary of making scientific inferences from our own particular circumstances. The notion of geocentrism—that the earth was at the center of the cosmos, with all other bodies orbiting about it—held sway. As early as the fourth century BC, most educated Greeks thought that the earth was a sphere at the center of the universe (Fraser 2006, 14). Plato wrote about it in a chapter in the *Republic,* providing a (partial) theological (or mystical) interpretation of the cosmos as the "Spindle of Necessity", turned by the three Fates and attended by Sirens. A more mathematical description (or explanation?)—known only from secondary sources—was given by Eudoxus of Cnidus, a student of Plato, asserting that all heavenly phenomena followed uniform circular motion. Aristotle, Plato's most famous student, imparted a considerable degree of additional sophistication to this model, positing the spherical earth to be at the center of the universe, with as many as 56 transparent concentric spheres rotating about it, to which all other heavenly bodies are attached. Ptolemy standardized this system, modifying it to include deferents and epicycles, in the *Almagest,* a culmination of centuries of work by Greek and Babylonian astronomers. For more than 1,000 years this was the standard cosmological model, and came to be called the geocentric model.

There were apparently good empirical reasons undergirding the model. Most obviously, the sun and all other celestial objects appeared to daily revolve about the earth, each star circling back to its rising point each day. Less obviously, if the earth were not fixed at the center, then a shifting of the fixed stars should be observable via stellar parallax. No such parallax was observed until the nineteenth century (an observation indicating that the fixed stars were much further from earth than originally postulated by Greek astronomers). Finally, the constancy of the luminosity of Venus implied that it was always the same distance from earth, consistent with geocentricity and not heliocentricity (which implies that Venus's luminosity should vary over time).[1]

Yet the heliocentric model of Copernicus supplanted the geocentric model, albeit quite slowly. While the model had the advantage of explaining the retrograde motions of the planets, a novel property that intrigued the German Archbishop of Capua, Nikolaus von Schönberg, and prompted him to encourage Copernicus to publish his findings, it had a number of disadvantages that prevented its rapid acceptance. It was no easier to use nor more precise than the Ptolemaic geocentric model in its predictions of planetary motions. It offered no physical explanation for the motion of the earth, whereas the Aristotlean/Ptolemaic system naturally kept the earth fixed and the stars moving due to their respective massive and aetherial natures. No observational proof was available to support the model, and the absence of stellar parallax that it predicted had to be explained away by conferring vast distances upon the fixed stars, which would have to be absurdly huge, as Brahe inferred from measurement (Blair 1990). Furthermore, this anti-Copernican 'evidence' of a fixed earth was understood to be in accord with the principles of Holy Scripture (Gingerich and Voelkel 1998). It was only the cumulative weight of observation and analysis by Brahe, Kepler, Galileo, and Zupi, culminating in Newton's universal law of gravitation, that caused the scientific community to accept the heliocentric model.

Many regard this acceptance as the full-fledged birth of modern science. Certainly the shift in perspective from geocentrism to heliocentrism—the 'Copernican revolution'—had far-reaching implications both for and beyond science. It would take another hundred years for Cyrano de Bergerac, Bernard le Bouvier de Fontenelle, and Johann Goethe to confer upon Copernicus his role of usurper of earth's (and with it humankind's) 'privilege' of being at the center of the universe (Danielson 2001). In the twentieth century, the notion of a Copernican Principle, a term first coined by Bondi in the context of proposing a steady-state cosmology, exerted considerable influence upon the modern scientific agenda, including general relativity, cosmological models, probes of the cosmic microwave background, cosmic inflation, and string theory (Danielson 2009). Its primary modern deployment has been to develop geometric models to describe cosmological observations that have no privileged point in space—every point can be regarded as the center of the space. In this sense every point is typical or mediocre, implying that on sufficiently large scales the universe is both isotropic (it looks the same in all directions) and homogenous (appearing to be the same from all places). Quite recently the Copernican Principle has been recast as a 'principle of mediocrity', whose implications are the main philosophical motivator for considering the multiverse (Vilenkin 2007, 163).

MEDIOCRITY

The principle of mediocrity is the assertion that "if an item is drawn at random from one of several sets or categories, it's likelier to come from the most numerous category than from any one of the less numerous categories" (Kukla 2009, 20). Applied as a scientific principle, it implies that any given phenomenon should be regarded as a typical specimen drawn from some larger sample, rather than assuming it exceptional, special, or privileged. In hindsight, geocentrism's error was to disregard this principle by assuming a special location for the earth within the universe. Applied more broadly, one can say that the mediocrity principle implies that there is nothing unusual about the structure or evolution of the solar system, earth, human beings, life, nations, or galaxies.

So should all natural phenomena be understood from the perspective of mediocrity? A few moments' thought indicates that considerable caution is required here, for there are many natural phenomena that indeed seem to be quite atypical. These include the shapes of the continents, the appearance of rare diseases, the complete left-handedness of biomolecules, the complete left-handedness of the weak interactions, the total eclipse of the sun by the moon, the universality of the speed of light, the geographical origin of human evolution, the galactic habitable zone, specific trajectories in chaotic systems, the linearity of quantum mechanics, the platypus, the 20 amino acids and 5 sugars that compose the chemical foundation of life, the values of the fundamental constants, the shapes of pottery in archaeological digs, the properties of water, and the beginning of the universe. Each of the various scientific disciplines that deal with these phenomena finds their own niche for how to interpret such particularity. Sometimes they point to an underlying general principle, as in plate tectonics' providing an explanation for the appearance of continents (though not for the particular broad features of continental shapes). Sometimes they are coincidental outliers from a broad sample, as in the moon's ability to totally eclipse the sun—out of the sample of all possible relative sun/planet/moon sizes and orbits, one might expect that a small subset will consist of the right configuration of moon size/orbit to exactly cover the disc of the sun during an eclipse; our earth-moon system happens to be in this rare subset. Sometimes they are evidently intrinsic to the structure of the universe—so far all variants of quantum mechanics that supplant the linear character of its equations with some other structure have been ruled out by experiment, yet from a mathematical viewpoint linear equations are as logically admissible as any other kind.

Particularity would then seem to be the flip side of typicality, and its often subtly learned lessons are ignored at our own scientific peril. In most of the sciences we can draw upon considerable empirical support in finding the appropriate epistemic equilibrium for particularity/mediocrity. Statistical information can be gleaned as appropriate in a variety of ways, yielding better guidance for judging the special cases.

However, in the case of the cosmos as a whole, we are presented with a special problem. We have only one observable sample of the phenomenon in question, that is, the universe itself. Our application of the principle of mediocrity—the Copernican Principle—therefore requires considerable care. For it is both a principle via which our observations are interpreted and it is an assertion that is inferred from our observations. Appropriately applied, it has brought considerable explanatory coherence to cosmological observations, allowing us to understand cosmic uniformity: the homogeneity and isotropy of the cosmos at large scales. Yet those same observations have shown us the limitations of cosmic mediocrity, since they have indicated that our universe has a developmental history that began 13.7 billion years ago at what is called the Big Bang. In strong contrast to scientific expectations of the early twentieth century, not all times in cosmic history are typical. Furthermore, it appears that we now live at an atypical cosmic 'moment', in which the energy responsible for cosmic acceleration is comparable to that of normal mass/energy, something that occurs only once.[2] Even further, all directly detected forms of mass/energy (such as matter from the periodic table) compose a small fraction (less than 5 percent) of the total mass/energy content of the universe, with dark matter and dark energy (whose existence is only known inferentially from astronomical observation) making up the remainder. More perplexing still is the fact that our best model of modern cosmology can agree with observations only when its parameters are very precisely adjusted. Small changes in these parameters result in significant disagreement with experiment/observation, indicating the large-scale properties of our universe are in an apparent state of very delicate balance (Carr and Rees 1979). The aforementioned cosmic uniformity and the measured magnitude of the amount of dark energy (or the cosmological constant) are examples of this. The latter quantity disagrees with its expected theoretical value by factor of 10^{120}. This discrepancy can be ameliorated, but only at the price of finely adjusting various parameters to 120-decimal-place precision. Such "cosmic fine-tuning" is a situation regarded by many as the most embarrassing problem in theoretical physics (Susskind 2005).

COSMIC PARTICULARITY

Evidently our place in the cosmos has some atypical features. But what of our universe itself? Is it a typical specimen among all possible universes?

Even this appears not to be the case. The origin, laws, development, and present-day configuration of our cosmos all appear to be somewhat delicately balanced for life to exist. While of course we can only live in a cosmos that permits (at least our kind of) life to exist, what is intriguing is that such biophilic (or life-admitting) universes form a rather tiny subset out of the vast range of possible universes we might contemplate.[3] These biophilic features, enumerated elsewhere,[4] include the relative masses of the stable elementary particles (the electron, proton, and neutron), the rate of expansion of the universe, and the relative strengths of the forces of nature (electromagnetism, gravity, weak, and strong). For example, if the neutron were just 0.2 percent lighter (heavier), then all protons would decay (fuse to neutrons) in the early universe, leaving in either case no hydrogen, no subsequent formation of water, and thus no life. In this sense, out of all possible universes that differ only in the magnitude of the neutron mass, none can support life except for those whose neutron masses differ from that of our own within 0.2 percent. There are a considerable number of such narrow biophilic selection effects, cumulating to render our universe rather finely tuned to support life.

Note that it could have been otherwise. Had the biophilic selection window for the neutron mass (and many other constants of nature) been, say, 20 percent, we would have concluded the existence of life does not require any further consideration; insofar as the existence of life was concerned, our universe would have been regarded as ordinary. Yet observation indicates that biophilic selection effects are quite sharp, given known physical law.

Turning to the foundations of physical laws, there are (at least) two other striking examples of particularity. One is the direction of time.[5] While the laws of classical physics are temporally neutral—for example, Newton's laws of mechanics are the same if the direction of time is reversed—the observed situation for our universe is quite different. The most direct example of this is psychological—we remember the past, but anticipate the future. This is but one of several temporal arrows. In electromagnetism, wave phenomena always ripple outward from a charged source via so-called retarded wave forms; yet the equations permit a time-reversal of this situation in which advanced wave forms ripple inward to a charged source. In gravitation, black holes absorb all forms of mass-energy whereas no (known) time-reversed "white holes" exist that emit all forms of energy. In subatomic physics, Kaons (and B-mesons) decay in a manner preferentially distinct from their corresponding antiparticles, a phenomenon to which one can ascribe another arrow of time. Biologically, evolutionary adaption yields an increasing complexity to all forms of life with increasing time. In contrast to this, the total thermodynamic entropy of any system with its environment is always observed to increase with time, a situation referred to as the 2nd law of thermodynamics. And, as noted earlier, the observation that we live in an expanding universe originating from the Big Bang indicates a cosmological arrow of time.

The origin of and relationship between these temporal arrows is rather mysterious. Our known laws of classical physics—including general relativity, which implies that time and space have a degree of fungibility—do not yield the temporal directionality that we observe in a broad variety of phenomena. In the context of thermodynamics, the resolution of this conundrum is to confer a low entropy to the initial state of our universe (Boltzmann 1897), so that the progenitors of today's isolated subsystems were far out of thermodynamic equilibrium in the past. Yet this is hardly a resolution: first, because such an emergent thermodynamic arrow of time relies on an unexplained initial state of very low entropy; and second, because it does not explain why our cosmos apparently has only one such thermodynamic arrow instead of many. There is no logical inconsistency in living in a cosmos in which entropy increases with time in certain regions while in others entropy decreases with time[6]—yet we do not observe this situation.

Another temporal arrow is associated with quantum mechanics, itself a striking example of the particularity of our universe. Quantum mechanics relies on two laws of evolution (von Neumann (1932)). The first is given by the time-reversible Schrödinger equation (or its relativistic generalizations), which specifies how a quantum state (or wavefunction) of an isolated (sub) system evolves with time. The second law specifies that when an (ideal) measurement is carried out on the (sub)system at a given time, the quantum state is projected (the wavefunction collapses) onto the measurement outcome; this law is not time-reversible, yielding the quantum-mechanical arrow of time. However, the striking particularity of quantum mechanics primarily resides not in this temporal arrow but rather in the linearity of the time-reversible first law, which implies that sums of quantum states obey the same evolution equation as each of their constituents. This feature underlies all of the counterintuitive phenomena we associate with quantum theory, including wave-particle duality, entanglement (or nonsignaling superluminal correlations), teleportation, tunneling, and more. Despite a host of other possible time-reversible equations one might conceive, this one appears to be operative throughout the entire cosmos, having applications in early universe physics, stellar interiors, stellar spectra, and all atomic and subatomic phenomena. Phenomenologically viable modifications of this basic linearity have proved to be enormously difficult to construct, far more so than the corresponding exercise in general relativity (in which generalizations of the basic equations are much easier to obtain), which in part explains why attempts to unite quantum theory with gravity have for the most part focused on generalizing gravitational physics in some way while leaving the basic structure of quantum theory intact. Furthermore, the presumed emergence of the everyday world of classical physics we see is via the rather poorly understood measurement law, whose deeper foundations continue to elude us.

COSMOCENTRICITY

We therefore appear to be in a rather special cosmocentric situation, similar to the geocentric one of old. We have a set of fairly sharp biophilic selection effects, a set of correlated temporal arrows operative throughout the observable universe, and a rather particular quantum foundation to physical law whose relationship to our everyday classical world is still not understood. What should we infer from this particularity?

This is a meta-question, one going beyond the bounds of standard scientific inquiry. While it is tempting to retreat into standard Copernicanism, we must exercise due caution. Like our geocentric predecessors, we observe ourselves to be in a setting with a variety of puzzling particularities, in a cosmos with atypical features. But unlike them, we have only one observable sample. A response to this latter comment might be that our geocentric predecessors were likewise technologically limited to a single observable sample, being unable to contemplate the advent of the telescope and other measuring devices. However, a counterresponse is that we have a considerably larger accumulation of scientific knowledge on which to build: we can state with very high degrees of confidence that perpetuum mobiles cannot be constructed, that we are nearing the maximum energy threshold for which we can carry out subatomic physics experiments, and that the speed of light will necessarily limit our observational knowledge of the cosmos. While we would be foolish to preclude further scientific innovation, it is likewise very reasonable to expect that we are indeed empirically restricted to a single cosmic sample. We therefore need to understand our atypical cosmos on its own terms, appealing to the most internally coherent explanation. We should not be surprised if this journey takes us beyond scientific inquiry into theological territory. Do the distinctive features of our cosmos signify that we are part of a Creation? Can we reasonably regard its particularity as evidence of the existence of a Mind?

Is contemplating Mind a destination for rational thought, or is it a form of magical thinking relying on wishes and fairy tales? The scientific instinct is to avoid the latter and join empiricism with reason to discern the structure and details of the underlying mechanism for any given phenomenon. At the least, the assumption of Mind is generally regarded as an unproductive scientific strategy,[7] in part because of the lack of sound criteria for implementing this assumption along with its accompanying teleological notions. Instead, an ecbatic[8] approach to understanding reality is taken, one that regards explanations of observable effects to be profitable and constructive only when they hypothesize natural causes: specific mechanisms, not magic or indeterminate miracles. This technically agnostic approach— often referred to as "methodological naturalism"—is not necessarily in contradiction with a theistic perspective that seeks to discern meaning behind mechanism (Polkinghorne 1991). It can almost be regarded as the definition of scientific inquiry. Yet for many, methodological naturalism is not far from philosophical naturalism, the assertion that nature is all there is and all basic truths are truths of nature (Audi 1996). Although the former does not logically imply the latter, it has been argued that the lack of any sound criteria for discerning supernatural processes, combined with the empirical success of methodological naturalism, indicates that philosophical naturalism is the only stance for any reasonable person to adopt (Forrest 2000).

This stance sets theism in opposition to naturalism, seemingly ruling out the possibility of ever inferring on any empirical grounds that there is a Mind underlying existence. It appears to force one to regard the existence of the cosmos, with its life, minds, and particularity emerging via fully undirected reductionist processes (Crick 1994). Theists can reasonably respond that such strong rationalism does not address either the problem of consciousness (and its indefatigable resistance to be understood reductionistically) or the question of existence itself: what it is that puts "fire into the equations of physics and makes a universe for them to describe"? (Stephen Hawking, as quoted in Ferguson 1995, 143). They could furthermore argue that understanding existence in the context of Mind (or perhaps Logos) offers the advantage of offering a single coherent framework for understanding objective reality and subjective experience.[9]

OPTIONS FOR PARTICULARITY

Returning to the question of explaining our own cosmos and its particularity, what options remain? There appear to be four broad categories. While not fully mutually exclusive, they are distinct enough to merit separate consideration.

One is that the cosmos is simply some random realization of a set of potential situations. This stance would appear to be at odds with both naturalism and theism. It amounts to abandoning any hope of finding a deeper meaning for, or explanation of, existence. Its chief flaw is an inability to come to grips with the high degree of intelligibility and rational transparency that the scientific enterprise has led us toward, offering no explanation for the source of cosmic potentiality. If the function of rational thought is to make sense out of apparent absurdity, then this avenue must be rejected.

A second possibility would be some form of cosmic necessity. The idea is that there is some kind of unified mathematical/scientific description of the cosmos.[10] Such an ultimate theory would have no adjustable parameters (or perhaps one) from which all other observable constants of nature and phenomena could be deduced. Ideally, such a theory would be logically necessary, with all other competitors having some kind of fatal mathematical flaw hidden in their core.

Certainly the trend of scientific thinking has been along such lines of unification. Newton's law of gravitation unified celestial physics with terrestrial physics. The periodic table provided an understanding of all matter as being composed of fewer than 92 stable basic elements. Maxwell's theory of electromagnetism unified the apparently disparate phenomena of magnetism and electricity. The quark model afforded a description of a plethora of subatomic particles as bound states of only a few distinct constituents. A unified theory of all forces and particles would seem to be a natural next step—perhaps even an endpoint—along such lines.

The problem with this idea is that it so far has failed to deliver the goods. The simplest models unifying the strong interaction with the electroweak interaction (the latter itself a unification of radioactive (or weak) forces with electromagnetism) have been ruled out by experiment, and there is no indication at this point in time from the highest energy experiments at the Large Hadron Collider (LHC) of any further unification (Bechtle et al. 2012). While this current disappointing situation might be rectified once the LHC is refurbished to reach its maximum collision energy, the hope of any unified theory to explain the particularity of the cosmos has yet to be realized, despite considerable effort on the part of the theoretical physics community. The values of the various masses of the elementary subatomic particles (electrons, muons, quarks, etc.) and the relative strengths of the various forces have yet to yield to any deeper mathematical description. Indeed, the basic equations of the standard model (the current description of all known forces and particles) admit solutions for pretty much any values of these quantities, with only their sharp biophilic character perhaps hinting at some deeper explanation.

Another possibility, noted earlier, is the inference that the cosmos is a creation. The idea here is that Mind is the source of the particularity of our universe, weakening cosmic necessity to cosmic intent, even if that intent is only the whim of a cosmic trickster. To go beyond such a capricious view of existence will in part require drawing upon the various religious traditions, taking seriously the possibility that the Mind has revealed itself to (at least) those parts of creation able to contemplate it. Thematic in all religions, the assertion that Mind is fundamentally the root of existence perhaps reaches its zenith in the gospel of John, which states that that "In the beginning was the Word, and the Word was with God, and the Word was God" (John 1:1).[11] It is Logos—Logic, Reason, Meaning, Thought, Word—that is the ultimate source of all things. Intelligence is the source of our intelligible existence. The particularity of the universe is regarded as indicative of the choices and intentions of a Creator. This perspective is open to most forms of further scientific discovery along the lines of unification; indeed such unification is expected from the Logos of the cosmos. However, the range of inquiry is broadened. Religious experience is accorded considerable merit, playing a role crudely analogous to that of experimentation in science. Theological reflection—appropriately informed by scientific inquiry—becomes the primary approach for understanding the Logos.

Much more detail can be given in terms of how theological inquiry proceeds. But the guiding paradigmatic framework is that there is a Telos underlying existence. Within Christian thought there is a further assertion that the hallmark of this Telos is neither indifference nor chicanery, but rather love. The Creator is particular not only in the choices made for the existence of the cosmos, but also in the interactions with the cosmos, and specifically with those beings having the capacity to have a relationship. This particularity is focused most sharply in the person of Jesus Christ, regarded as a unique concrete manifestation of the loving Mind behind existence.

The fourth possibility—the most recent trend for explaining cosmic particularity—is to apply the principle of mediocrity to its fullest extent, regarding our observable cosmos as a small part of a much larger structure. Somewhat analogous to the manner in which a molecule of oxygen is one specimen out of a very large number of air molecules on earth, this perspective posits that our universe is one out of a very large number of existing universes. The entire collection of universes is called the multiverse. The principle of mediocrity implies that just as a molecule selected at random from the air should be typical—most of its various properties (speed, mass, rotation) should fall within a standard deviation of all the allowed possibilities—most of the measureable properties of our universe should likewise be typical, falling within a standard deviation of the allowed possibilities.

MULTIVERSE MOTIVATIONS

The multiverse approach is driven by a desire to avoid cosmocentrism and any special assumptions that go along with it. This philosophical perspective is not without indirect scientific support. Cosmic fine-tuning and sharp biophilic selection are regarded as indirect evidence for this approach,[12] and mechanisms exist for generating different kinds of universes with differing laws, structures, and initial conditions. The best known of these come from string theory and cosmic inflation. String theory—originally thought to be the ultimate realization of cosmic necessity insofar as it aspires to unify all of particle physics and gravity into a grand quantum theory—was shown a number of years ago to have as many as 10^{500} kinds of low energy states, each corresponding to a universe with its own particular properties and features.[13] Cosmic inflation posits the existence of a state of matter known as a false vacuum, whose key property is that its interactions with gravity cause the universe to double in size every 10^{-34} seconds; if this happens only for 10^{-32} seconds, the universe will increase in size by a factor of 2^{100} or 10^{30}. Cosmological uniformity, instead of resulting from a delicate balance of conditions at the Big Bang that would otherwise yield chaotic heterogeneity, emerges because of the strong smoothing effects of such a rapid expansion (Guth 1981, 1998; Linde 1982). Some (as yet unknown) mechanism cuts off this rapid expansion very shortly after 10^{-32} seconds, yielding a fragment of "true vacuum" that contains our observable universe, which then expands as per observation. The ubiquity of the false vacuum implies that bubbles of true vacuum perpetually percolate to generate universes bearing some resemblance to our own. Extensions of this idea allow different directions for the arrow of time in different universes, yielding on average no temporal direction for the multiverse as a whole (Carroll and Chen 2004; Hartle 2013).

However, for the multiverse paradigm to be efficacious, these different universes must physically exist;[14] they cannot be hypothetical entities. Were the latter situation to be the case, then our atypical cosmos would be a single instantiation of a plethora of possibilities, and one would need a mechanism for selecting it. However, if all possible universes are physically instantiated, then our universe must be one of these, and its special features occur simply because all possible features occur somewhere/when in the multiverse.

Superficially, the multiverse paradigm has a certain attraction. It is a-teleological, in the tradition of methodological naturalism. It offers a research program of sorts, insofar as one can attempt to construct probability measures for the various parameters of fundamental physics and develop methods for counting universes. Could it be the most intellectually coherent response to cosmocentrism? And to what extent is it compatible with other approaches, particularly theism?

THE PROBLEMS OF PLENITUDE

Consider first the intellectual coherence of the multiverse paradigm. As with theism, to make judgments on this score will entail going beyond normal scientific inquiry, since direct observation of universes outside of our own is by definition impossible. We will have to conform our modes of inquiry and rational judgment to the subject of investigation.

Yet herein lies the first (and still unresolved) problem. How does one go beyond normal scientific inquiry and still do science? How and in what way should the rules of science be changed to accommodate the multiverse paradigm? We are far from traditional scientific falsification here, since we cannot empirically go beyond our single observable sample. Unlike theological inquiry, the multiverse paradigm does not have any significant tradition or resources for exploring the phenomenon at hand. The rules of the game are at best ill-defined (Ellis 2006). While mathematics can provide some guidelines, a physical theory necessitates some schema for identifying mathematical constructs with physical entities. Of course theoretical physicists do this all the time, and one might expect this approach to develop its own traditions and rules organically, as has historically been the case elsewhere. However, direct empirical guidance is not and never will be forthcoming, breaking with all known scientific approaches.

But what of the principle of mediocrity? Perhaps we can determine if the various features of our cosmos are typical among the spectrum of possibilities statistically spread across the ensemble of allowed universes (Vilenkin 2007, 163). We could perhaps test our own 'typicality' by observing (apparently) bio-irrelevant constants of nature, such as neutrino masses or lifetimes of unstable quarks and see if they attain typical values within a standard deviation of some normal distribution. However, this poses the question of what criteria should be used to construct such normal distributions. A failed test could be regarded as a flaw in the choice of normal distribution as opposed to a refutation of the multiverse paradigm.

Using the principle of mediocrity to temper the rules of science to admit radical changes in what is regarded as a legitimate foundation for a physical theory is by no means a straightforward or simple exercise. A key element of the problem here is the implicit (sometimes explicit) use of unbounded resources. In the case of cosmic inflation, the false vacuum has this property. However, all multiverse approaches—cosmological natural selection (Smolin 1992), many-worlds quantum theory, the string landscape among them—rely upon an arbitrarily large amount of resources (matter, energy, time, space) to implement their various universe-generating mechanisms. Even a simple universe that is infinite in spatial extent has this property. Such unboundedness entails an implicit reliance on an arbitrary number of outcomes in order to ensure that our observed situation is, if not typical, then actualized.

Yet multiverse practitioners proceed, extrapolating the laws and properties of our observed cosmos to spatiotemporal regions well beyond empirical scrutiny. In so doing, two key troublesome features have emerged: Boltzmann Brains[15] and Rampant Duplication (Ellis and Brundrit 1979; Tegmark 2003). The community of cosmologists regards the former as a significant problem, whereas the latter is regarded as a quirky idiosyncrasy.

I contend otherwise.[16] Rampant duplication is a situation that arises as a consequence of repeatedly generating universes via some mechanism. Without some cap on the generation mechanism, eventually a universe emerges containing an exact copy of a given individual, since DNA has a finite number of configurations (Ellis and Brundrit 1979). For example, in a multiverse of infinite spatial extent (where the various universes are regions demarcated by the distance light could have travelled since the Big Bang—the so-called Hubble volume), each human on earth has a doppelganger a finite distance away. Rough estimates of this distance imply that the closest copy of any given human $10^{10^{29}}$ away from here, far larger than the 10^{26} meter radius of the Hubble volume of our universe. Any given subsystem in our Hubble volume will be replicated, including our entire Hubble volume itself, which in the previous model is estimated to be $10^{10^{118}}$ metres away.

This is strange enough. Even stranger, such duplicates will appear arbitrarily often in all possible environments, as will all possible variants of duplicates that are allowed by the laws of physics governing the multiverse under consideration. This means that any given physical subsystem traverses all logically admissible trajectories arbitrarily many times. In other words, anything that can happen, does happen, and that infinitely often. Rather than being a quirky trait, this feature undermines the purported explanatory power of the multiverse paradigm, since it is no longer clear how to reliably infer a general law based on particular outcomes, or rule out unlikely possibilities on the basis of chance. Quantum-mechanical probabilities are no longer given by the absolute squares of quantum amplitudes (the Born rule becomes nonoperative), and the replacement prescription for making predictions is uncertain (Page 2010a). For any given experiment—social, psychological, biological, physical—all logically allowed experimental outcomes occur somewhere in the multiverse, undermining the principle of induction.[17]

The Boltzmann brain problem takes this a step further, introducing the possibility that our actual experience is no longer trustworthy. The situation here is that complex structures have a very small but nonzero probability to emerge by chance from a system in equilibrium, given sufficient time and resources. One such structure is a human brain, with a full set of memories and perceptions. It could emerge as a thermal fluctuation from an infinite universe in thermal equilibrium or as a quantum fluctuation from the vacuum. If one uses the presence of observers to delineate certain subsets of the multiverse (e.g., biophilic universes), then there is no reason to regard such Boltzmann brains as having 'observer status' inferior to that of any other human (or bio-brain).

The problem is that of the two, bio-brains (the kind we think we have) are the atypical complex structure. The best of our scientific knowledge indicates that they have a long evolutionary history and will one day become extinct in an expanding cosmos. We perceive ourselves to have brains of this type. Boltzmann brains, on the other hand, while extraordinarily rare to emerge from a fluctuation (either per unit time or per unit volume), never become extinct, since such fluctuations occur repeatedly often in time. Hence over any sufficiently long eon within a segment of the multiverse, the number of Boltzmann brains will vastly dominate the number of bio-brains, contrary to the original motivation of the paradigm. There has been a hefty amount of research directed toward this problem, but no solution is forthcoming at present.[18]

MULTIVERSE THEOLOGY

What of the theistic implications of the multiverse? I have written elsewhere that the multiverse introduces a new theological question: why is there something rather than everything? (Mann, forthcoming). The multiverse paradigm entails production of a sufficiently large number of universes to ameliorate cosmocentricity. However, there is no clear bound on this number in any known multiverse theory; the number is arbitrarily large or effectively infinite. The implication of this, noted earlier in terms of rampant duplication, is anything that can happen actually does happen within the premises of a given multiverse model. Unlike all other scientific theories, in which the actual, physically instantiated, states of a system are a finite subset of the (sometimes infinitely) large potential states of the system, in the multiverse paradigm all potential states must be actualized in order to make use of the principle of mediocrity. This might appear to be overstating the case: only a finitely large number of universes (and hence a finite number of actualized states) need to exist for mediocrity to be effective. However, a complete theory would need to provide at least both an estimate and a rationale for such a finite number—and so far no multiverse theory has been forthcoming on this point. Furthermore, the nature of all such approaches to date—eternal inflation being a good example—is to perpetually generate universes. Once universe-generation is started, it is not clear how it stops (Mann 2005).

Yet we only observe a finite number of things in our universe: a finite number of particles, a finite number of choices, finite information content. If this observation is taken to mean that not everything exists, then we are left with the question as to why there is something rather than everything. The traditional puzzle of existence—why there is something rather than nothing—is turned on its head. Creatio-ex-Nihilo becomes Creatio-ex-Omnia.

If theism is regarded as a satisfactory answer to the nonexistence of nothing, it would also seem to be a satisfactory answer to the nonexistence of everything. For a God that chooses to create existent things is presumably a God that is selective in the choice of things to create.

Or is this the case? Why couldn't God create a multiverse? Why couldn't God create all possible states and situations? Indeed, by what criteria should God's creative power be limited? Who are we to say that God didn't create more than one universe? Or all possible universes, for that matter?

Such an argument is sometimes put forth by theists who wish to contend that the idea of a multiverse is not in conflict with traditional notions of God, particularly that of Christian theology (Page 2010b). I disagree. Positing a deity that creates everything is a radical departure from standard monotheism, particularly the Christian God. With some trepidation, we must take up Zophar's challenge to Job about probing the limits of the Almighty (Job 11:7). Of course God's characteristics are—by definition—without bound. Biblical language describes God as having wisdom without limit (omniscience), being the source of all power (omnipotence), as having love is too vast to be grasped (omnibenevolence). (See Psalm 147:5, Deuteronomy 3:23–25, and Romans 8:39, respectively.) In contrast to this, the creation is limited, subordinate to and dependent on God for its origin, existence, and fulfillment (Job 38:4–11).

The limitless character of the multiverse necessitates a radical revision of this picture, giving rise to new theological tensions between previously congenial aspects of God's character. For example, the extent to which a putatively elegant mathematical description of the multiverse is reflective of the intelligibility of God is undermined by the imbecilic generation of all conceivable universes resulting from God's unbounded power. Existing theological tensions are also amplified to unreasonable extremes. A case in point is the theodicy problem, in which God's love stands juxtaposed to the origin and presence of evil. It is difficult enough to deal with this situation in our own world—it would seem intolerable to complete all incidents of evil being repeated arbitrarily many times (Mann 2009).

The challenge of multiverse theology is that of actualizing all logically admissible outcomes within a given set of constraints while avoiding total absurdity. It is not clear that this challenge can successfully be met. One could imagine imposing some kind of theological limitation on the multiverse—perhaps in any given universe there is an upper bound on the amount of injustice or evil, by whatever metrics one might use to quantify such concepts. However, to do so is to deviate further from standard scientific practice by imposing direct theological limits on a scientific theory. Certainly all multiverse models to date make no use of this kind of constraint. Indeed, it undermines the original motivations for the multiverse, which is to generate universes by some a-teleological (random) process so as to impose the principle of mediocrity.

Another viewpoint involves reliance on the hope of finding some deeper principle underlying the multiverse, one that more fully reflects the existence, glory, and intelligibility of its Creator. This is perhaps the best argument for putting theism in consonance with the multiverse paradigm. It cannot be ruled out, and mathematical elegance historically has been a hallmark of fruitful scientific theories. However, this perspective is not without difficulty.

One difficulty is that the present state of the art is far from this situation. This point has been made by Susskind (2005, 125), who, concerning the generation of the multiverse via string theory, stated:

> String theorists watched with horror as a stupendous Landscape opened up with so many valleys that almost anything can be found somewhere in it. The theory also exhibited a nasty tendency to produce Rube Goldberg machines. In searching the Landscape for the Standard Model, the constructions became unpleasantly complicated. More and more "moving parts" had to be introduced to account for all the requirements, and by now it seems that no realistic model would pass muster with the American Society of Engineers—not for elegance in any case.

The same assessment might be given of pretty much any multiverse approach at present. Eternal inflation generates bubbles of true vacuum containing standard model physics, but it also chaotically generates other bubbles containing pretty much any other logically admissible physics as well (Linde 1986; Aguirre and Gratton 2002). Cosmological natural selection relies on some basic principles of population biology as applied to universe-generation through black hole production, but it does not give any guidance as to how the parameters of physical theory undergo variation as new universes are generated (Altenberg 2012), nor does it rule out any other radically different laws of physics.

A deeper issue is that there is an unbounded cornucopia of elegant mathematical principles, any of which might be regarded as the foundation for a multiverse. This point has been made by Tegmark (2003) in the context of the so-called level-4 multiverse, in which every mathematically possible universe is physically instantiated, the only constraint being internal self-consistency. If this approach is pursued, then by what criterion could we say that one mathematical structure is more or less elegant than another?

A third difficulty is sociological: a driving force behind the multiverse paradigm is nontheism. Ever since the days of the steady-state model of cosmology (Hoyle 1948), the scientific community has been disturbed by theistic implications of the transient nature of our cosmos (Smith 1992). Much effort has been expended to develop models that continue the universe prior to the Big Bang,[19] restoring a measure of temporal permanence to the cosmos. The exceptional biophilic character of our cosmos has intensified this drive to avoid its theistic implications, as evidenced by recent popular literature on the subject.[20] Proponents of a theistically consonant multiverse should at the least recognize the deep nontheistic motivations behind the subject if they intend to articulate a credible contrary manifesto.

Faith is the substance of things hoped for, the evidence of things not seen (Hebrews 11:1). If so, then it must remain an article of faith that the multiverse will bring forth some deeper principle of elegance. But such faith is resting on the hope that we have correctly carried out the gargantuan extrapolations from known physics into unknown territory, that we have correctly gleaned hints as to the way things really are from all possibilities, and that current marginal evidence against the multiverse (such as the value of the total density parameter or the lack of observed collisions that our universe has had with other vacuum bubbles) will go away.[21] This is hardly the hallmark of motivated belief. I suspect that the multiverse science will take on a very different form should some principle of elegance prove to be correct.

SUMMARY

It is the ongoing effort of both science and theology as to how to properly reconcile the particular with the general. Careful judgment is needed to discern those observations and experiences that point to general principles and concepts from those whose atypical character merit special consideration.

I have argued here that cosmocentricity—the observation that our observable universe is laced with a number of uncommon properties—represents perhaps the most extreme case in point. While the experience of geocentrism has prompted a phalanx of cosmologists and theorists in recent times to apply a modern version of Copernicanism to render our universe as one of many in an unboundedly large set of possibilities, I have pointed out that this approach comes with a high epistemic price. It is not obvious that it is either good theology or good science, though I suppose a counterargument to the latter assertion is that there are a sufficient number of number of reputable people working on the subject to make it so. Yet the scientific problems are quite formidable, involving very large extensions of present-day knowledge into completely unknown realms with no forthcoming empirical corroboration. While it is the job of theoretical physics to take adventurous leaps, that subdiscipline is likewise charged with the task of responsibly critiquing their form and destination. Can mediocrity lead us astray? If it failed, how would we know?

Theologically, the multiverse does generate a new question for reflection (Mann 2009). Why is there something rather than everything? While we cannot ascribe the same measure of confidence to the assertion that everything does not exist that we can to the assertion that something does exist, our knowledge of modern-day cosmology suggests that it is quite reasonable to give serious consideration to the specificity of the cosmos from a theological perspective. This includes not only our special biophilic physical laws, but also the correlation of the arrows of time, the foundations of quantum theory, and other particularities of known scientific findings. The mediocrity perspective needs to be balanced by a sound measure of contemplation of the remarkably special features of our universe.

Our particular cosmic setting is indeed rather puzzling. At what point is it appropriate to take such particularity at face value, employing alternatives to the mediocrity principle in order to achieve a deeper understanding of the way things are?[22]

NOTES

1. Today we know that the increase in the apparent size of Venus caused by its varying distance from Earth is compensated by the loss of light caused by its phases.
2. See Armendariz-Picon, Mukhanov, and Steinhardt (2000, 2001), Bonvin, Caprini, and Durrer (2006), Akhoury, Garfinkle, and Saotome (2011), and Leonard et al. (2011).
3. For a philosophical analysis of these issues, see Leslie (1989).
4. For a comprehensive discussion, see Barrow and Tipler (1988).
5. An early exposition of this idea is given in Eddington (1928).
6. See Schulman (1999, 2001, 2005).
7. See, e.g., Wilczek (2007).
8. For a more complete description of the meaning of this term, see Mann (2005).
9. For a full discussion of this kind of broad integration, see Murphy and Ellis (1996).
10. An early attempt along these lines was given by Georgi and Glashow (1974).
11. All Bible quotations are from the New International Version.
12. See Ferguson (1995) for further discussion.
13. For a readable introduction, see Bousso and Polchinski (2004).
14. For a collection of articles that address this subject from a variety of perspectives, see Carr (2007).
15. These are named after Ludwig Boltzmann, who first introduced the issue (Boltzman 1895). See also Rees (1997, 221), Albrecht and Sorbo (2004), and Page (2008). For a recent nonspecialist discussion, see Overbye (2008).
16. This is a point I have made elsewhere; see Mann (2009).
17. For a discussion of the principle of induction in the context of the multiverse, see Leslie (2009).

18. For a review of the situation, see Overbye (2008).
19. For an overview of recent efforts along these lines, see Ananthaswamy (2008).
20. A few examples are Susskind (2005), Weinberg (1994), Stenger (1988), Dawkins (2008), Greene (2011), and Hawking and Mlodinow (2010).
21. These points were made by Ellis (2005).
22. I am grateful to Klaas Kraay for the invitation to speak at the God and the Multiverse Workshop (Ryerson University, February 15–16, 2013) and to its participants for their critique of the ideas presented here.

REFERENCES

Aguirre, A., and S. Gratton. 2002. "Steady-State Eternal Inflation." *Physical Review D* 65: 083507.

Akhoury, R., D. Garfinkle, and R. Saotome. 2011. "Gravitational Collapse of k-Essence." *Journal of High Energy Physics* 1104: 096.

Albrecht, A., and L. Sorbo. 2004. "Can the Universe Afford Inflation?" *Physical Review D* 70: 063528.

Altenberg, L. 2012. "Implications of the Reduction Principle for Cosmological Natural Selection." arXiv:1302.1293.

Ananthaswamy, A. 2008. "Did Our Cosmos Exist Before the Big Bang?" *New Scientist* 2686: 32–35.

Armendariz-Picon, C., V. Mukhanov, and P. Steinhardt. 2000. "Dynamical Solution to the Problem of a Small Cosmological Constant and Late-Time Cosmic Acceleration." *Physical Review Letters* 85: 4438–41.

Armendariz-Picon, C., V. Mukhanov, and P. Steinhart. 2001. "Essentials of k-Essence." *Physical Review D* 63: 103510.

Audi, R. 1996. "Naturalism." In *The Encyclopedia of Philosophy* (Supplement), edited by D. Borchert, 372–73. New York: Macmillan.

Barrow, J. D., and F. Tipler. 1988. *The Anthropic Cosmological Principle*. Oxford: Oxford University Press.

Bechtle, P., et al. 2012. "Constrained Supersymmetry after Two Years of LHC Data: A Global View with Fittino." arXiv: 1204.4199.

Blair, A. 1990. "Tycho Brahe's Critique of Copernicus and the Copernican System." *Journal of the History of Ideas* 51: 355–77.

Boltzmann, L. 1895. "On Certain Questions in the Theory of Gases." *Nature* 51: 413–15.

Boltzmann, L. 1897. "Zu Hrn. Zermelos Abhandlung über die Mechanische Erklärung Irreversiler Vorgänge." *Annalen der Physik* 60: 392–98.

Bonvin, C., C. Caprini, and R. Durrer. 2006. "No-go Theorem for k-Essence Dark Energy." *Physical Review Letters* 97: 081303.

Bousso, R., and J. Polchinski. 2004. "The String Theory Landscape." *Scientific American* 291: 79–87.

Carr, B., ed. 2007. *Universe or Multiverse?* Cambridge: Cambridge University Press.

Carr, B. J., and M. J. Rees. 1979. "The Anthropic Principle and the Structure of the Physical World." *Nature* 278: 605–12.

Carroll, S., and J. Chen. 2004. "Spontaneous Inflation and the Origin of the Arrow in Time." arXiv: hep-th/0410270.

Crick, F. 1994. *The Astonishing Hypothesis: The Scientific Search for the Soul*. New York: Touchstone.

Danielson, D. 2001. "The Great Copernican Cliché." *American Journal of Physics* 69: 1029–35.

Danielson, D. 2009. "The Bones of Copernicus." *American Scientist* 97: 50–57.
Dawkins, R. 2008. *The God Delusion*. New York: Houghton Mifflin Harcourt.
Eddington, A. 1928. *The Nature of the Physical World*. Cambridge: Cambridge University Press.
Ellis, G. F. R. 2005. "Physics Ain't What It Used to Be." *Nature* 438: 739–40.
Ellis, G. F. R. 2006. "Issues in the Philosophy of Cosmology." In *Philosophy of Physics*, edited by J. Butterfield and J. Earman, 1183–286. Amsterdam: Elsevier.
Ellis, G. F. R., and G. B. Brundrit. 1979. "Life in the Infinite Universe." *Quarterly Journal of the Royal Astronomical Society* 20: 37–41.
Ferguson, K. 1995. *The Fire in the Equations: Science, Religion, and the Search for God*. Grand Rapids, MI: Eerdmans.
Forrest, B. 2000. "Methodological Naturalism and Philosophical Naturalism: Clarifying the Connection." *Philo* 3: 7–29.
Fraser, C. G. 2006. *The Cosmos: A Historical Perspective*. Westport, CT: Greenwood.
Georgi, H., and S. L. Glashow. 1974. "Unity of All Elementary-Particle Forces." *Physical Review Letters* 32: 438–41.
Gingerich, O., and J. R. Voelkel. 1998. "Tycho Brahe's Copernican Campaign." *Journal of the History of Astronomy* 29: 1–34.
Greene, B. 2011. *The Hidden Reality: Parallel Universes and the Deep Laws of the Cosmos*. New York: Knopf.
Guth, A. 1981. "Inflationary Universe: A Possible Solution to the Horizon and Flatness Problems." *Physics Review D* 23: 347–56.
Guth, A. 1998. *The Inflationary Universe*. New York: Perseus.
Hartle, J. B. 2013. "The Quantum Mechanical Arrows of Time." arXiv: 1301.2844.
Hawking, S., and L. Mlodinow. 2010. *The Grand Design*. London: Bantam.
Hoyle, F. 1948. "A New Model for the Expanding Universe." *Monthly Notices of the Royal Astronomical Society* 108: 372.
Kukla, A. 2009. *Extraterrestrials: A Philosophical Perspective*. Lanham: Lexington Books.
Leonard, D., J. Ziprick, G. Kunstatter, and R. B. Mann. 2011. "Gravitational Collapse of K-Essence Matter in Painlevé-Gullstrand Coordinates." *Journal of High Energy Physics* 1110: 028.
Leslie, J. 1989. *Universes*. New York: Routledge.
Leslie, J. 2009. *Infinite Minds: A Philosophical Cosmology*. Oxford: Oxford University Press.
Linde, A. D. 1982. "A New Inflationary Universe Scenario: A Possible Solution of the Horizon, Flatness, Homogeneity, Isotropy and Primordial Monopole Problems." *Physics Letters B* 108: 389–93.
Linde, A. D. 1986. "Eternally Existing Self-Reproducing Chaotic Inflationary Universe." *Physics Letters B* 175: 395–400.
Mann, R. B. 2005. "Inconstant Multiverse." *Perspectives on Science and Christian Faith* 57: 302–10.
Mann, R. B. 2009. "The Puzzle of Existence." *Perspectives on Science and Christian Faith* 61: 139–50.
Mann, R. B. Forthcoming. "Thinking of Everything." In *Irreconcilable Differences*, edited by J. Robinson. London: Pickering and Chatto.
Murphy, N., and G. F. R. Ellis. 1996. *On the Moral Nature of the Universe: Theology, Cosmology, and Ethics*. Minneapolis, MN: Augsburg Fortress.
Overbye, D. 2008. "Big Brain Theory: Have Cosmologists Lost Theirs?" *New York Times,* January 15.
Page, D. 2008. "Is Our Universe Likely to Decay within 20 Billion Years?" *Physical Review D* 78: 063535.
Page, D. 2010a. "Born's Rule Is Insufficient in a Large Universe." arXiv: 1003.2419.

44   *Robert B. Mann*

Page, D. 2010b. "Does God So Love the Multiverse?" In *Science and Religion in Dialogue*, vol. 1, edited by M. Stewart, 380–95. Oxford: Blackwell.
Polkinghorne, J. 1991. *Reason and Reality: The Relationship between Science and Theology*. London: SPCK.
Rees, M. J. 1997. *Before the Beginning: Our Universe and Others*. New York: Simon and Schuster.
Schulman, L. S. 1999. "Opposite Thermodynamic Arrows of Time." *Physical Review Letters* 83: 5419.
Schulman, L. S. 2001. "Resolution of Causal Paradoxes Arising from Opposing Thermodynamic Arrows of Time." *Physics Letters A* 280: 239–45.
Schulman, L. S. 2005. "Two-Way Thermodynamics: Could It Really Happen?" *Entropy* 7: 208–20.
Smith, Q. 1992. "A Big Bang Cosmological Argument For God's Nonexistence." *Faith and Philosophy* 9: 217–37.
Smolin, L. 1992. "Did the Universe Evolve?" *Classical and Quantum Gravity* 9: 173–91.
Stenger, V. 1988. *Not by Design: The Origin of the Universe*. New York: Prometheus Books.
Susskind, L. 2005. *The Cosmic Landscape: String Theory and the Illusion of Intelligent Design*. New York: Little, Brown.
Tegmark, M. 2003. "Parallel Universes." *Scientific American* 288: 41–51.
Vilenkin, A. 2007. "Anthropic Predictions: The Case of the Cosmological Constant." In *Universe or Multiverse?*, edited by B. Carr, 163–80. Cambridge: Cambridge University Press.
von Neumann, J. 1932. *Mathematische Grundlagen der Quantenmachanik*. Berlin: Springer.
Weinberg, S. 1994. *Dreams of a Final Theory: The Scientist's Search for the Ultimate Laws of Nature*. New York: Vintage.
Wilczek, F. 2007. "A Model of Anthropic Reasoning: The Dark to Ordinary Matter Ratio." In *Universe or Multiverse?*, edited by B. Carr, 151–62. Cambridge: Cambridge University Press.

2   The Everett Multiverse and God

Don N. Page

Science looks for the simplest hypotheses to explain observations. I begin by setting out some simplicity-based considerations favoring the Everett multiverse of many 'worlds'. I then consider what I think may be an even simpler principle of why reality is as it is; namely, that *the actual world is the best possible world*. From this assumption, I sketch what I call the *'Optimal' Argument for the Existence of* God. Briefly, this argument holds that the sufferings in our universe might not be consistent with its being alone the best possible world, but that the total world could be the best possible if it includes an omnipotent, omniscient, omnibenevolent God who experiences great value in creating and knowing a universe with great mathematical elegance, even though such a universe has suffering. Further considerations show that such a being might well create a multiverse. If God exists, He seems loathe to violate elegant laws of physics that He has chosen to use in His creation, such as Maxwell's equations for electromagnetism or Einstein's equations of general relativity for gravity within their classical domains of applicability, even if their violation could greatly reduce human suffering (e.g., from falls). If indeed God is similarly—and justifiably—loathe to violate quantum unitarity (even though such violations by judicious collapses of the wavefunction could greatly reduce human suffering by always choosing only favorable outcomes), the resulting unitary evolution would lead to an Everett multiverse. There are, then, both scientific and theological considerations favoring this view.

SIMPLICITY

Humans and especially scientists tend to favor the simplest explanations for observations, and this motivates principles such as Occam's razor. For example, physicists are currently searching for a mathematically elegant complete theory of the universe, with one candidate partial theory being superstring/M theory.[1]

One may attempt to quantify the plausibility of competing candidate theories by a Bayesian analysis in which one initially weights theories by their prior probabilities (logically before considering observations) but then multiplies these prior probabilities by what is somewhat misleadingly called the likelihoods of the theories, which are the conditional probabilities of one's particular observation, given each of the theories (assuming, in turn, that each theory is correct). When this product of the prior probability and the likelihood for each theory is normalized by dividing by the sum of the products for all theories, one gets the posterior probability for the theory—the probability that it is correct given the observation. Of course, this is a highly idealized procedure, since we do not have even one single complete plausible theory enabling us to calculate the probability of any particular observation. Also, the prior probabilities appear to be unavoidably subjective.[2]

Even if we make the subjective choice to use Occam's razor and assign higher prior probabilities to simpler theories, it is subjective which theories are simpler. Furthermore, even if one did produce an ordering of theories, say with the nth theory being simpler than the $(n+1)$th for each positive integer n, there would still be the subjectivity of how to choose a set of prior probabilities that decrease with n. (A simple choice one could make for a countably infinite set of theories ordered in increasing complexity would be prior probabilities that are $1/2^n$, but there are infinitely many other algorithms for constructing an infinite sequence of decreasing real nonnegative numbers that sum to one as probabilities should.)

Although it is subjective how to assign the prior probabilities, and we do not yet know how to calculate the likelihoods for any plausible complete theory (one that would predict the probabilities of all possible observations), there does seem to be a trade-off between the priors and the likelihoods. For example, the simplest theory I can think of (to which I might assign a prior probability of 1/2) is the theory that nothing concrete exists. However, this gives zero conditional probability for one's observation, so its posterior probability is zero. The second simplest theory I can think of is that all possible observations exist with equal probability (to which I might assign a prior probability of 1/4), but since presumably there are an infinite number of logically possible observations, the likelihood of that theory, the probability of one's particular observation, would be $1/\infty$, so this theory would also have a posterior probability of zero.

At the other extreme, the maximum-likelihood theory would be the one that gives a probability of one to one's observation. Since the probabilities for observations given by a theory must sum to one, this theory must give zero probability for all other observations, so it is an extreme solipsistic theory, saying not only that no one else's observations exist, but also that no other observations exist for the one in question at any other time. We cannot logically rule out such a maximum-likelihood theory, but I suspect that one's observation is sufficiently complex that any complete theory predicting it and only it with positive probability would have to be sufficiently complex that it would be reasonable (though still subjective) to assign that theory an extremely low prior probability. Then it is highly plausible that other complete theories exist (though presently unknown in complete detail) that would reasonably be assigned sufficiently higher prior probabilities such that the products of those priors with the (necessarily lower, and probably quite a lot lower) likelihoods of those theories would still be higher than the product of the prior and the unit likelihood of the extreme solipsistic maximum-likelihood theory.

THE EVERETT MULTIVERSE

One can apply these principles to the Everett version of quantum theory. This version says that when a quantum choice arises between different physically allowed outcomes, not just one outcome occurs, but all of them that are allowed occur. However, any single observation can include an awareness of just one of the outcome possibilities, so different observations occur in what are called different Everett 'worlds'.[3]

Different outcomes can have different measures of existence, however, so the different observations that can occur in different Everett 'worlds' generally have different probabilities, which are the normalized measures for the observations. Just as different observations of different observers within a single Everett 'world' can mutually exist, so do the different observations in different Everett 'worlds'. This means that the probabilities of the different observations should not be interpreted as if they were something like propensities for potentialities to be converted to actualities, with only one potential observation being actualized, but rather the probabilities should be interpreted as normalized measures for the existence of many different actual observations. Not only do both you and I both have observations now in one single Everett 'world', but also there are many other observations of different versions of us, and of very many other observers as well, in very many other Everett 'worlds' that all lie within one quantum state for a universe in the single actual world (assuming that the Everett picture is correct).

Qualitatively, the idea of many different observations in many different Everett 'worlds' is just an extension of the idea of many different observations by many different observers at one time in one Everett 'world', and also of the idea of many different observations by one observer at many different times. In fact, in quantum cosmology it is natural to think of different times as different Everett 'worlds', so if you believe in the existence of many different times, you already believe in one form of many Everett 'worlds'.

Although there are many different competing versions of quantum theory, perhaps the most straightforward competing realist version is a collapse version that postulates that on certain occasions the quantum state collapses to just a single Everett 'world' at that time, so that for each time and each observer, there is just one observation. (This version still has the many observations of many different observers at many different times, so it only partially avoids the problems of plenitude that bother Mann in Chapter 1 of this volume.) For such a version of quantum theory to be made complete (e.g., to be able to give the normalized measures for all observations), it would need to give a precise specification of when the quantum state collapses and to what single 'world' it collapses to. No such specification has been given, so at present, collapse versions of quantum theory are much more incomplete than the Everett version, which avoids the problem by avoiding any collapses of the quantum state.

Furthermore, no experimental evidence has ever been found for any collapses of the quantum state. On the other hand, there are many cases in which the quantum state has been found to evolve without collapsing through a superposition of what one would naturally call different Everett 'worlds' that then come together again to form a coherent superposition that leads to observable effects in agreement with no collapse but in clear disagreement with the postulate that a collapse occurred during the experiment.

Now, a proponent of a collapse version of quantum theory could correctly object that we do not know of any actual observations (say in the sense of conscious perceptions or sentient experiences) occurring in the different 'worlds' of the quantum superposition that are found to come together coherently (without collapsing). The cases in which such different 'worlds' combine again without collapsing so far have involved far fewer (though still a large number of) atoms than those in conscious observers that we know about. So therefore such a proponent of collapse could claim that the collapse only occurs when the complexity gets greater than what is possible in the experiments showing no collapse, or perhaps when consciousness is involved. However, since the Everett version with no collapse agrees with all experiments so far, it seems to me unnecessary to postulate that there is some collapse that occurs beyond the present experimental ability to be detected.

Furthermore, it seems highly plausible that any precise specification of the collapse that would not be ruled out by experiments showing no collapse in many relatively simple cases would be much more complex than the Everett version with no collapse. Therefore, if one indeed assigns simpler theories higher prior probabilities, it would seem that a complete theory incorporating the Everett version of quantum theory would have much higher prior probability than collapse versions.

There is the flip side that the Everett version of quantum theory would lead to a far larger number of observations than a collapse version (at least in a sufficiently small universe; if the universe is sufficiently large, which might be a consequence of the simplest complete collapse theories that fit observations, it is not at all clear that a collapse version of quantum theory would solve this problem). This means that the normalized probabilities for most observations would tend to be much lower in complete Everett theories than in complete collapse theories. This is how I would interpret the problems of plenitude that Mann raises in Chapter 1 of this volume.

I do believe that this diluting of the normalized probabilities of observations is a real cost of any multiverse theory (or even of any multiobservation theory, such as a theory with many observers, or even of just one observer making many observations over different times). If the complete theory predicts equal probabilities for all the observations that one might suppose are physically allowed, I do suspect that the normalized probabilities will be diluted so much that when one uses the probability of one's own observation that is predicted by this theory as the likelihood of the theory (even after multiplying by the rather high prior probability I would assign to such a simple complete theory that assigns equal probabilities to a huge set of possible observations) one would get a posterior probability that is far lower than a better complete theory that does not assign equal probabilities to all the physically allowed observations. Robert Mann's objections to multiverses seem to apply mainly to this simple multiverse theory, and I agree that they seem quite telling against it.

However, the hope is to find a more sophisticated multiverse theory (no doubt at some cost in complexity that would lead it to be assigned a lower prior probability) that would give sufficiently higher probabilities to our observations that whoever used the probability of his or hers as the likelihood of the theory would get the product of the prior probability and the likelihood much higher than this simple multiverse theory and perhaps even higher than what one would get for any single-universe theory. The idea is that although the simplest multiverse theory that weights equally all possible observations would lead to a posterior probability (the product of the prior probability and the likelihood, normalized by dividing by the sum of such products over all complete theories) that is much less than a good single-universe theory, the greater simplicity (and hence higher prior probability) of a good multiverse theory could outweigh the lower likelihood that it has from the greater dilution of the normalized probabilities of observations in that multiverse theory that has more observations.

Note that when we postulate evaluating the posterior probability of a complete theory as the normalized product of the prior probability and the likelihood that is given as the probability that the theory gives for the observation of the one testing it, there is no need for the observation to be of the entire multiverse in a multiverse theory, or even of the entire universe in a single-universe theory. So it does not count against the scientific testability of multiverse theories that other parts of them are not observable by us. It is sufficient that one can evaluate the product of the prior probability and the likelihood of the theory and be able to compare that with the product for all other theories that do not lead to a much smaller product. Therefore, in principle, a complete multiverse theory is just as testable as a single-universe theory. Considering multiverse theories is by no means giving up science or the importance of observations.

Now of course it must be admitted that we do not yet have any plausible complete multiverse theory to test. We also do not have any plausible complete single-universe theory to test.[4] It will require a lot of work, and the postulation of new principles, to find good candidates, not only for the dynamical laws of physics, but also for the quantum state and for the rules that give the probabilities of observations from the quantum state. Such candidate complete theories should avoid having the probabilities of observations dominated by Boltzmann brains[5] or by other types of observations qualitatively much different from ours.

It may be overly optimistic to hope that while our civilization lasts we will be able to find a good complete theory. Perhaps it is too hard for us humans to find, particularly given the limits on our future lifetime within this universe. But since science has made such enormous progress in recent centuries, it would certainly seem to be a worthwhile goal to try to make progress toward such a complete theory. Of course it is too early to tell, but I suspect that we shall find scientifically that something like the Everett version of quantum theory, along with perhaps an even broader multiverse that includes different values of the effective coupling constants of physics, works better that competitors to explain our observations.

Therefore, if Occam's razor is used (in the form of assigning higher prior probabilities to simpler theories), along with the principle of observation (to use the normalized probability that the complete theory assigns to one's observation as the likelihood that one would assign to the theory), one can in science seek after the complete theory that gives the highest posterior probability, the highest normalized product of the prior and the likelihood. This would in a sense be the simplest theory that fits the observations, but there is the trade-off between simplicity (affecting the prior) and the fit to observations (given by the likelihood). Neither the simplicity nor likelihood would be maximized separately.

So neither maximal simplicity nor maximum likelihood seems plausible in the sense of giving a theory with nearly the maximum posterior probability. In physics, one might imagine that a theory with high posterior probability is a fairly elegant mathematical theory, such as the Everett version of quantum theory along with a rather simple dynamical theory, the way superstring/M theory is hoped to be.[6]

THE 'OPTIMAL' ARGUMENT FOR THE EXISTENCE OF GOD

However, even if such a theory were correct, it might still be considered to leave unexplained why this rather simple, but not maximally simple, theory is correct rather than an even simpler theory. One possibility is that there is an even simpler principle of why the world is as it is. Although we humans have a very poor grasp on it (and so usually find a mathematical theory easier to calculate with), one principle that could conceivably be a simpler explanation is the following hypothesis about the actual world: *the actual world is the best possible world.*

Let me explore my dim understanding of this hypothesis and consider what it might plausibly be taken to imply, eventually leading to the question of whether or not there is an Everett multiverse. I am not claiming that the steps in this argument are by any means logically complete, but I am offering them as a very crude sketch for what seems to me as if it might be a plausible picture of the world.

By hypothesizing that the actual world is the best possible world, I mean that in some sense it maximizes value. I take the value that is maximized to be the intrinsic value of conscious or sentient experiences. A painting may have instrumental value in eliciting positive feelings in a viewer, but I take the real ultimate value to be the pleasure or happiness or joy (or whatever positive word one might use) of the sentient experience of the viewer. That is, I am taking all other values just to be means to this end: if there are no sentient experiences, there is really no value at all. Assuming that the painting itself has no conscious awareness, I do not ascribe to it any intrinsic value. Similarly, the laws of physics are presumably not sentient, but I do believe that they have great instrumental value in leading to pleasure in conscious experiences.

So what I mean by intrinsic value excludes entities without consciousness (which can only have instrumental value). However, I do not assume that entities with consciousness have only intrinsic value, the value of their own conscious experiences. Humans can be highly instrumental in bringing about positive conscious experiences in others. Indeed, usually when we speak of a good person, we are not speaking of the intrinsic value of his or her own sentient experiences, but of the way his or her actions lead others to have positive conscious experiences. We do not say Hitler was a good man even if he experienced happy thoughts while dancing a jig during World War II. The suffering he caused is sufficient reason to label him as evil, whatever his own experiences may have been.

But when I consider the total goodness of the world, I do mean intrinsic value—pleasure minus suffering, happiness minus unhappiness, joy minus agony—of all conscious experiences that occur. I am supposing that in our actual world, this total goodness is the greatest that is logically possible.

I am ignoring the problem that perhaps there is no maximum to intrinsic value. Infinity is a problem for very many theories about the world, so that in such cases one could not make progress if one stopped just as soon as one faced this admittedly severe problem that does eventually need to be addressed. Here I simply assume that there is a maximum. I am also assuming that all sentient experiences can be put onto a single scale of intrinsic value, which is perhaps analogous to the assumption that all economic goods can be priced on a single scale of currency. Of course, I do not know the precise intrinsic value for most sentient experiences (and not even of the one I am presently having, though I have a strong sense that currently it is positive). However, I see no evidence against the assumption that each sentient experience does have a precise value (or at least a value relative to other experiences; I doubt that there is a preferred absolute scale to value).

Now, I have personally experienced a lot of happiness in my life, and I have seen many other people appear to be happy, so I do think that our universe does have considerable positive intrinsic value. But there is also much suffering and unhappiness, so it does not seem very plausible that the intrinsic happiness within this universe *alone* is the maximum logically possible.

However, besides the mixed happiness that our universe contains, our universe seems to exhibit a very high degree of mathematical elegance and beauty. For example, even without the complete unification of the laws of physics, we can say that to an excellent approximation electromagnetism is described by Maxwell's equation, which can be written in the simple form $*d*dA = J$, and gravity is described by Einstein's equation, which can be written even more simply as $G = T$.[7]

Humans can partially appreciate some of this elegance, so it does seem to increase the intrinsic value of the universe. On the other hand, most of us (even we diehard physicists) would surely be happier if disasters, diseases, and cruelty were eliminated from the universe, even at the cost of less mathematical elegance and beauty for the laws of physics.

But if one considers the possibility that there may be a Being outside the universe that has an enormous appreciation for the mathematical elegance of the universe (no doubt far greater than our dim appreciation of it), and perhaps an even greater appreciation if that Being knew He were the Creator of the universe, then it seems that the hypothesis that the actual world is the best possible world might plausibly lead one to believe that there is such a Being outside the universe who experiences tremendous value in having much greater knowledge and appreciation for the mathematical elegance of the universe than we creatures within the universe can have.[8] I call this line of thought the *'Optimal' Argument for the Existence of God.*

One might by further considerations of simplicity postulate that such a Being not only has very great knowledge of the universe but actually has all possible knowledge, so that such a Being is omniscient (Swinburne 2004). Considerations of simplicity might also suggest that such a Being is not only capable of creating our universe but also can do anything logically possible that His nature desires to do, so that such a Being is omnipotent. And under our original assumption that the actual world is the best possible world, it seems plausible to assume that such a Being is all-good or omnibenevolent. Of course, I am not claiming that I have any logical proof from exclusively uncontroversial premises for the existence of such a Being or of these properties of omniscience, omnipotence, and omnibenevolence, but they do seem to me not unreasonable assumptions to make if one accepts the simple hypothesis that the actual world is the best possible world. The hypothesis that the actual world is the best possible world certainly appears to me to be more consistent with a world with an omniscient, omnipotent, omnibenevolent Being (which I shall henceforth call God, since these are the main properties traditionally assigned to God), than with a world in which our universe stands alone, with its mathematical elegance not fully appreciated.

In particular, if there were no God who is a Person (at least in the sense of having conscious experiences), and if our universe (perhaps a multiverse) were the only entity with any conscious experiences within it, then it would seem that such a world could be better if all the conscious experiences in our universe were perfectly happy. This might not be consistent with the actual laws of our universe, but surely it is logically possible and could be the case if the laws were suitably different from what they actually are. Therefore, under the assumption of no personal God, our observations would seem to be incompatible with the hypothesis that the world is the best possible. (Of course, atheists rarely do make the hypothesis that the world is the best possible, though even Richard Dawkins has said, "The world and the universe is an extremely beautiful place, and the more we understand about it the more beautiful does it appear".[9]) Despite my appreciation as a physicist for the beautiful laws of physics, I myself would much prefer to give up these laws whenever necessary to prevent cancer in people I know, or earthquakes in places such as Haiti.

On the other hand, if there is an all-knowing God who is completely aware of the entire universe He creates and fully appreciates the mathematical beauty of its laws of physics that He uses in His creation, this omniscient conscious awareness could have enormous value and help make the entire world, including God and His own sentient experiences, the best possible world. (This would, of course, not be the only feature that makes the entire world the best possible; it presumably is better that it have created sentient beings rather than just a universe created to maximize elegance but without sentient beings.) Cancer and earthquakes may be logical consequences of these laws of physics. God Himself may grieve over the evils that are a consequence of the laws of physics that give Him even much greater joy, but there may be this inevitable trade-off.

Note that I am not saying that God is constrained by the laws of physics in the sense that they are any external limitation on His happiness. But if He had chosen to eliminate the evil, the laws of physics that He would have had to use to do that would have been different from what they actually are. Quite possibly they would have had to be less elegant and beautiful to Him than the actual laws He did choose, and such less-orderly laws may well have made Him much less happy, so that the total value of the world could well have been reduced.

One might complain that the choices God make appear selfish, placing His happiness above that of His creatures. But if God's choice really did maximize the total happiness of the world, then I see it as quite justified. Who are we to complain when God does what He pleases, especially if what He does actually maximizes the total intrinsic value of the world?

What I think is wrong about most human selfishness, including much of what I see in myself, is that it places a higher value on one's own happiness than on that of others. Far too often we act to increase our happiness by a certain amount, whereas if we had acted differently, other people's happiness could have increased manyfold. For example, if those of us who are well-off in developed nations contributed even just 10 percent of our income to help those less well-off in poorer nations, our personal happiness might go down by a tiny fraction, but I strongly suspect it would go up by a much larger fraction for those helped. I am not saying that it is wrong to enjoy happy experiences and to seek to increase them, so long as doing that does not cause a greater decrease in happiness in others, but I am saying that very often it would be better to sacrifice a bit of one's personal happiness to give a greater increase in the happiness of others.

Therefore, if it is indeed true, as I postulate, that God gets tremendous satisfaction from the elegant laws of physics that He creates, it would not be right to expect Him to give that up to the degree that would be required for a much smaller happiness that might be afforded His creatures if they were spared from evils like cancer and earthquakes. God has created this universe to lead to much human joy, and I suspect that He has indeed sacrificed much personal happiness to achieve that. In particular, I believe that Jesus greatly reduced the happiness He otherwise would have had in Heaven by coming to earth as a human and enduring the suffering and shame He did, especially when He was unjustly executed on the Cross. But I think it would be unfair for us to expect God to give up more happiness than we would gain by some action of His, such as perhaps changing His usual laws of physics to prevent cancer and earthquakes.

THE THEOLOGICAL ARGUMENT FOR AN EVERETT MULTIVERSE

In the Everett 'many worlds' version of quantum theory, a person is continually branching into many copies (each copy in a different Everett 'world'). Even with exactly the same genes and previous experiences (the same 'nature' and 'nurture'), the outcomes in the different Everett 'worlds' will be different. In ours, Hitler was an evil monster. But I suspect that in most Everett 'worlds' with the same early 'nature' and 'nurture' for Hitler, he was not nearly so evil. (Of course, there is the flip side: in most Everett 'worlds', Mother Teresa presumably also did not turn out so good as she did in ours.)

I believe that it is a consequence of the laws of physics that when a person is faced with a moral choice, in some Everett 'worlds' in which that choice is made, an evil choice is made, one that reduces the total happiness of the conscious beings in that Everett 'world'. There will also be Everett 'worlds' in which a good choice is made, which increases total happiness in that 'world'. (One might postulate that Jesus was an exception, choosing to incarnate Himself with no quantum amplitude to make any evil choices.)

If God had chosen to use sufficiently different laws of physics (perhaps even just to the extent of collapsing the quantum state appropriately as needed), I think He could have eliminated not only all natural evil, but also all human evil. However, that might actually have decreased the total happiness of the entire world by reducing His own happiness greatly at using less elegant laws of physics. (Admittedly, I do not see a logical necessity in this, so besides the controversial assumption that the world is the best possible, I am still assuming that if a God exists, He has a certain nature that would greatly prefer most of the time using elegant laws of physics.)

This consideration leads into the theological argument for an Everett multiverse. Henceforth I shall be assuming the existence of a God who loves mathematical elegance as well as other sentient beings, so the remainder of this essay does not aim to support the *'Optimal' Argument for the Existence of God*. Instead, it aims to explain certain features of the world that would result from such a being's existence.

As explained earlier, it seems highly plausible that leaving the quantum state uncollapsed is mathematically much more elegant and beautiful than collapsing it as would seem to be needed to avoid superpositions of macroscopically different situations (different 'Everett worlds'). (A lot of information would need to be specified to give a complete theory that includes collapse, which would apparently make the theory very complex, whereas elegant theories are relatively simple.) This, in fact, is the main reason I personally favor the uncollapsed Everett hypothesis about the quantum state. In my *'Optimal' Argument for the Existence of God*, I am assuming that a suitable omniscient God would greatly appreciate the elegance of the universe and prefer a more elegant version (such as Everett's) over a less elegant one, such as one with 'random' collapses of the quantum state.

But if one conjectured that it did not reduce God's happiness very much for Him to collapse the quantum state, it would seem that He would do so in ways to increase creaturely happiness under this new conjecture that it would not cost God more than the gain to us (an alternative to what I actually believe), so then it would seem that God would indeed collapse the quantum state in such ways to increase creaturely happiness, which apparently could be enormously increased by suitable collapses of the quantum state. Of course, one cannot rule out some other reason why God might choose to collapse the quantum state in ways that do not greatly increase creaturely happiness over that in our actual world, but it would certainly seem mysterious to me why an omnibenevolent God would not choose collapses to increase creaturely happiness greatly if that would be at less cost to His own happiness. The fact that He has apparently not done so suggests that there is a good reason why He has not, such as for maintaining the high level of pleasure He gets from very elegant laws of physics that do not have collapses of the quantum state.

Another alternative is that conceivably there is an elegant way for the quantum state to be collapsed, and God uses and greatly appreciates that elegant way, even though it does not increase creaturely happiness nearly so much as what seems logically possible from a suitably different set of collapses of the quantum state to avoid creaturely suffering and maximize creaturely happiness. For example, suppose that somehow there were some elegant set of discrete times at which the quantum state were to be collapsed, as well as an elegant discrete set of basis states at each of those times into which the quantum state could collapse. (It is not obvious how to make an elegant choice for either of these, but for the sake of argument let us suppose God has made some particular choices for them.) Then one might make the *ad hoc* postulate that at each of the collapse situations, God takes the two basis states that give the largest absolute value of the inner product with the actual quantum state before the collapse (with the basis state giving the largest value labeled 0 and the basis state giving the next largest value 1; it would be a set of zero cases, which we can ignore, in which there is a tie for the largest two values), and then God collapses the quantum state onto whichever one of those two basis states corresponded to the next binary digit of π.

This at least is a definite algorithm for which of the discrete basis states to collapse the quantum state onto at each of the discrete times, though it sets aside the question of how to choose the basis states and the times. The algorithm is fairly simple, being described by the single final sentence of the previous paragraph, so in some sense it might be regarded as fairly elegant, though to me is certainly does not look so elegant as simply leaving the quantum state uncollapsed, because it does require more information and more *ad hoc* specifications. Furthermore, once one completes the algorithm by giving a specification of the discrete times for the collapses and of the basis states at those times, the algorithm would presumably be even more complicated and less elegant. Therefore, even though it seems logically possible that God could collapse the quantum state in an elegant way (which most likely would not much increase the ratio of creaturely happiness to sadness over that in the Everett version of quantum theory with no collapses, since this algorithm does not seem to relate directly to whether the results are happy or sad, so in this way it seems consistent with our observations), it does seem to run the danger of making the universe less elegant and less pleasurable for God.

Thus it seems simpler to me to focus on the two extreme cases: (1) God collapses the quantum state to maximize creaturely happiness; and (2) God does not collapse the quantum state at all. Since (1) certainly seems contrary to our observations, (2) appears to me much more plausible. It might maximize total happiness once God's appreciation of the elegance of the universe is included.

Now of course I cannot rule out intermediate cases, such as the one described previously in which God does use a fairly elegant way to collapse the quantum state. Another alternative that I find much more plausible is that most of the time God does not collapse the quantum state, but on certain very special occasions, such as the Resurrection of Jesus Christ, and perhaps in certain other miracles, He does choose to do so for reasons that override His usual desire for elegance. However, it seems to me that most of the time, God does not choose to collapse the quantum state in a way that obviously appears to have much moral significance for us creatures within the universe, so when applying theological as well as philosophical and scientific considerations to our observations, it seems most consistent with my assumptions and with our observations that God rarely if ever collapses the quantum state.

In summary, I see not only philosophical and scientific but also theological reasons for an Everett multiverse, for the reality of different quantum outcomes to choices that the universe faces. These reasons are based upon the assumptions I have made in my *'Optimal' Argument for the Existence of God,* that the world is the best possible at least partially because God exists and has great appreciation for very elegant laws of physics that He creates. There are presumably other values as well, such as the intrinsic value of the sentient experiences in the universe God lovingly creates (rather than just creating a universe that is mathematically elegant but devoid of sentient beings), and perhaps even the intrinsic value of God's experience of knowing us as sentient beings He has created to have fellowship with Him.

Of course, as the Book of Job emphasizes, it is very difficult for us finite beings to understand why we are here or the motives of a Creator under the assumption that we were indeed created by God. However, if we have indeed been created in the image of God, we may have some dim appreciation of some of the things that God might value, such as elegance and simplicity. It then seems plausible to conjecture that God may value this elegance (among other things) to such a degree that He faithfully uses the same elegant laws over and over again, and that doing so does maximize the total value of the world, despite the evils and sufferings that result. Therefore, I continue to hold the tentative hypothesis that God has created our universe to make the total actual world the best possible.

NOTES

1. Presently, superstring/M theory is not complete, in part because it does not give the quantum state or the rules for getting the probabilities of observations from the quantum state. A so-called 'Theory of Everything' or TOE that just gives the dynamical laws is certainly not a complete theory for the universe. The dynamical laws would have to be augmented with a specification of the quantum state and by rules for extracting the probabilities of observations from the quantum state in order to qualify as a complete theory.
2. The correct objective prior probabilities would be one for the complete correct theory of the actual world and zero for all other theories. But since none of us knows the correct theory, we are forced to assign subjective prior probabilities to different complete theories for different logically possible worlds. To say that prior probabilities are subjective, however, is not to say that all

are equally justified. Some subjective prior probabilities are surely more plausible than others. However, what is considered plausible is also unavoidably subjective.

3. The Everett 'worlds' are different components of the quantum state that all occur within one world of all that exists. Since the Everett 'worlds' are not what philosophers call 'possible worlds' (possible total states of affairs), I shall put the Everett 'worlds' in single quotes.

4. I deem the simple theory that nothing concrete exists (which gives zero probability for our observations since it has no observations), and the simple theory that all logically possible observations exist with equal probability (which also gives zero probability for our observations, since the probability of any particular observation would be $1/\infty$) to be implausible, since they have zero posterior probability. I likewise deem the extreme solipsistic theory that just my present observation exists (which has the maximum likelihood of one but is so complex that I would assign it an extremely low prior probability) to be implausible.

5. Boltzmann brains are hypothetical observers that appear momentarily from thermal or other quantum fluctuations, which would be expected to have observations much more disordered than ours, so that if they dominated the normalized probabilities for observations, their contribution would dilute the probabilities for ours and hence reduce the likelihood of the theory.

6. Superstring/M theory is elegant in having a simple mathematical structure with very little arbitrariness. For example, it has absolutely no adjustable free parameters. However, no one knows a complete version of even just the dynamical part of this theory, to which one would need to add the quantum state and rules for getting probabilities of observations from it to give a complete theory.

7. For the experts, here * is the Hodge dual, d is the exterior derivative, A is the electromagnetic potential one-form, J is the current density one-form, G is the Einstein tensor for the curvature of spacetime, and T is the stress-energy tensor of matter. I am taking a particular choice of units that sets the speed of light, the permittivity of the vacuum, and 8 pi × the Newtonian gravitational constant all equal to 1.

8. Like all other arguments for the existence of God, this one makes assumptions that are not universally agreed upon, in this case especially the assumption that the world is the best possible. To me this assumption is more plausible than many of the assumptions in other arguments for the existence of God, but what appears plausible seems to be unavoidably subjective in a similar way that prior probabilities for different theories also seem to be. I should also note that I am not claiming that the appreciation of elegance that I am postulating for God is the only intrinsic value for Him, but if it is a great value for Him, that would appear to explain many observed features of our universe.

9. http://en.wikiquote.org/wiki/Richard_Dawkins.

REFERENCE

Swinburne, R. 2004. *The Existence of God.* 2nd ed. Oxford: Clarendon Press.

Theistic Multiverses
Details and Applications

3 The Multiverse
Separate Worlds, Branching, or Hyperspace? And What Implications Are There for Theism?

Peter Forrest

My purpose is to compare three basic versions of multiverse, that is, the many 'worlds', theory: Separate Worlds, Branching, and Hyperspace. There is, in addition, a variant on Branching, the Lattice Universe, as well as mixed theories. I shall reject Separate Worlds even though I assimilate merely possible worlds to those of Multiverse. The choice between Branching and Hyperspace is not so straightforward, although I judge that Hyperspace is to be preferred by theists. I then examine the consequences for theism, arguing that Hyperspace coheres better with theism than its rivals. In addition, Hyperspace supports the existence of something physical prior to God's act of creation, which I identify with the divine body. This 'something' could be described pretheoretically as without structure and indeterminate (the 'apeiron').

BACKGROUND

This essay is audience-specific in that I take for granted an approach to metaphysics that I shall expound but do not here have the space to defend. First I think that, *de facto,* metaphysics is speculative and that it should be a speculative scientific discipline. To say it should be a discipline is to submit that it should be conducted in as rigorous and clear fashion as is feasible, not with the mix of sensationalism and obscurity characteristic of much Continental philosophy. To describe it as scientific would be misleading without the 'speculative' qualification. For there is core science and speculative science, I say. And core science deserves to be called collective knowledge, with the authority this implies. Scientific progress, however, requires speculative research programs, whose Lakatosian 'hard cores' are not themselves core science (Lakatos 1970). By metaphysics I mean the science of the most fundamental truths, including the interpretation of core science itself.[1] (When I say that *de facto* it is speculative I am denying that it must always remain so, but at present we lack the consilience of different lines of argument required for it to achieve core science status.)

More specifically, I am operating within some assumptions that I devoutly hope will become collective knowledge, with the same status as core science. Currently I have to take these as speculative because many of my intellectual peers reject them. These assumptions, which make up Properly Anthropocentric Metaphysics (PAM), include the truth of core science, interpreted realistically. But they also include the denial that agency and consciousness occur in virtue of a more fundamental physical order. Hence PAM is compatible with physicalism only in some weak sense, and is incompatible with naturalism. One such compatible physicalism states that the nonphysical is correlated in a metaphysically noncontingent fashion with the physical. Another compatible physicalism is the denial that we have souls. In both these weak senses I consider physicalism a fairly probable piece of speculative metaphysics. PAM takes seriously the phenomenology of human freedom, while granting the relevance of brain science. Of importance to this essay is my assumption that the gravity of our choices is no illusion: a difficult choice is not like the toss of a coin.

MANY WORLD THEORIES

Universe-Fibers

The Universe, with an upper case 'U', is the sum of all that is physical. Multiverse (many worlds) asserts that the Universe is composed of many *universe-fibers,* as I call them, or *u-fibers,* for short. Informally, a u-fiber is roughly what eternalists (block theorists) used to think of as the whole universe, past present and future. I say 'roughly' because there is a version in which some u-fibers—maybe eventually all but one—terminate: that is, there is a space-like boundary beyond which the u-fiber does not extend. It is useful to think of an analogy in which the Universe is thought of as a book with an infinite number of pages—the u-fibers. I take the space dimension to be down to up and temporal dimension to be left to right.

It is not easy, however, to give a precise geometric definition of a u-fiber. We should not even assume that a u-fiber is of one temporal and three spatial dimensions. For String theorists typically consider 10 dimensions. Nor should we require u-fibers to be maximally connected parts of the Universe, for that would exclude Hyperspace. For the purposes of this essay I could operate with a disjunction: either String theory is correct and a u-fiber has 10 spatial dimensions, or it is false and every u-fiber has 3 spatial dimensions. In either case I take a u-fiber to have a (topological) manifold structure.[2] This stipulation stops us calling any branching structure a single u-fiber.[3] So, when I say a u-fiber branches (or splits or undergoes fission) I am saying that two or more u-fibers overlap (cf. Lewis 1983 on surviving fission[4]). In terms of the book analogy, consider a sheet that (like a sheet of paper) has a right hand edge, and suppose someone has glued onto that edge two sheets (each of which therefore has a left hand edge where it was glued).

I stipulate that this page that splits in two is not to be considered a single u-fiber but two u-fibers, which overlap to the left of the glue but are disjoint to the right. We may now imagine a most peculiar way of binding the pages in a book: start with a single sheet of paper, glue onto it two or more, and keep on repeating the process. This corresponds to one of the multiverse hypotheses, namely Branching. I am not, however, assuming that a u-fiber is nothing but a manifold. For the commonsense position, which might or might not be correct, would include its furniture, which we might think of as the print on the pages.

The preceding characterization requires that we stipulate the number of dimensions of a u-fiber. To avoid this stipulation we could use a definition that is not purely geometric but uses the concept of naturalness. It requires an assumption that I shall not rely on elsewhere, namely that there are no spatially extended atoms. In that case I characterize a u-fiber as having at least two dimensions (one of which is temporal) and being the result of a maximal *natural* (i.e., nonarbitrary), observer-independent, division of the sum of all ordinary particulars (such as events or material objects) into disjoint parts of at least one temporal and one spatial dimension. So the idea is that the multiverse is 'carved at the joints' into u-fibers, but that any finer division is arbitrary slicing—except *perhaps* the carving into successive temporal slices.[5] The restriction to ordinary particulars is intended to exclude not only universals but such things as laws of nature, which might not get divided up as required. It also excludes any nontemporal particulars there might be.

This way of characterizing the u-fibers permits Cross Fiber Extension, the thesis, for which I argue later, that all familiar particulars are extended across many u-fibers. On each page of the 'book' we might have a pattern of lines looking like a photograph of an interaction in a cloud chamber, or, if you prefer, a Feynman diagram. Ordinarily we would think of each of these lines as the history of a particle. On the thesis here being considered, however, the history of a particle extends some way through the book, and consists of different lines on the different pages. It is as if a diagram begins on page 30 and varies somewhat on each consecutive page until we get to page 70. Then Cross Fiber Extension tells us that an electron, say, belongs to many u-fibers and so is represented not just by one diagram but by many diagrams.

If there are extended atoms, then familiar particulars might be cross-world extended without having parts in distinct u-fibers, analogous to the way something with a temporal span might be said to endure rather than perdure. (Here I use the Mark Johnston / David Lewis Princeton terminology of endurance versus perdurance.[6]) If there are no extended atoms, then familiar particles are sums of, perhaps atomic, parts each of which is restricted to a single fiber.

Because Cross Fiber Extension asserts that familiar particulars extend across the u-fibers, we may introduce the idea of the lower case 'u' universes, consisting of many but not all u-fibers. *My universe as of now* is the join of the u-fibers to which the particles that make me up now belong.[7] This is either identical to or largely overlaps *your universe as of now*.[8] Generalizing, a universe is a bundle of u-fibers that share all or most of their familiar particulars. An important property of universes is that they can split even if the u-fibers do not branch, that is, even if no u-fibers overlap. That kind of fission occurs if the postfission universes containing just some of the u-fibers of the prefission universe. If such fission occurs, then your universe as of now is a proper part of your universe as of a second ago. Popular expositions of multiverse theories often talk of the universe splitting in this way, usually with the implication that we split with the universe. This is compatible with, but does not entail, Branching (see Forrest 2007).

Multiverse Hypotheses

All the hypotheses under consideration are of the multiverse genus. They, along with the genus itself, belong to speculative science-cum-metaphysics. And they all posit vastly more than a single u-fiber. On the hypothesis of Separate Worlds, there are many u-fibers with no direct spatial relations between them. The sheets of paper that we thought of as pages are not bound together, they are 'scattered'. Indeed they are not even spatially related. Consequently, the occupants of distinct u-fibers cannot share the same (spatial) location.[9] Branching, by contrast, is the hypothesis that the Universe is the sum of many overlapping u-fibers, subject to the rule that if location x belongs to both u-fiber v and u-fiber w, then so does location y if y is earlier than x. The 'earlier than' relation might be taken as the frame-invariant relation of being earlier in all frames, or on a dynamic theory of time, the relation that holds if y came to be present before x did. This constraint results in a tree-like structure in which we may think of a u-fiber splitting, but no two u-fibers fuse together.

Clearly there is a variant on Branching in which u-fibers *can* fuse together. I call this Lattice. On the book analogy, it is as if a single sheet is sometimes glued to two or more, at their right hand edge. When I compare hypotheses about the multiverse I include Lattice under Branching.

Hyperspace posits many u-fibers that occupy subspaces each of 4 (or for String theorists 10) dimensions in a Universe of $N + 1 = M + 4$ (or $M + 10$) dimensions, where M is the number of extra dimensions, and N the total number of spatial dimensions. This gives it the mathematical structure of a fiber bundle—the fibers are the u-fibers. If we considered the u-fibers to vary in n respects, then, we might expect M to equal n.[10]

On the book analogy, the u-fibers of hyperspace are distinct sheets that undergo neither fission nor fusion but are tightly bound together. Hyperspace itself can have a geometric structure. So if we are thinking of infinite sheets of paper we may think of Hyperspace as a three-dimensional space with the sheets being layers. Or to vary the image, consider cylindrical two-dimensional u-fibers with the infinite axis being temporal but space being circular. If we thought of these cylinders as differing only in radius, then they could all fit in as concentric layers in a three-dimensional Euclidean space with a missing central axis.[11] I note in passing the variant in which the division into u-fibers is relative to a relativistic frame of reference. Interested readers can work out the details, the rest may safely ignore it.

In addition to these three hypotheses, we might consider various mixed hypotheses. Thus there could be a tree in which some branches are 'fastigate', being themselves hyperspaces that branch, and there could be separated parts of the Universe not all of which are u-fibers but some are trees or hyperspaces. That would be like a book in several volumes.

FURTHER PRELIMINARIES

My comparison of the various multiverse hypotheses will be based on how well they provide the explanations that Multiverse is capable of. So before I make the comparison, I shall consider how Multiverse gives us a theory of agency, a theory of time, and a theory of physical probability. This will also provide a case for Multiverse itself. But first I shall make some preliminary points.

The Case for Cross Fiber Extension

The preceding definitions depend on a stipulated use of the term 'u-fiber' to mean something similar to what we used to think the Universe was. I now argue that our universe (as of whenever) is itself made of many but not all of the u-fibers. That is, Cross Fiber Extension holds. For consider an electron going through the twin slits of the idealized experiment. On Multiverse we explain its wave-like behavior by saying that in some u-fibers it goes through one slit and in others the other slit. That implies that both the electron and the observer have parts in many u-fibers. So our universe as of now is a bundle of many u-fibers.

Possible Worlds

There is one further distinction to bear in mind. The many u-fibers might be many possibilia or they might be many parts of the one actual world.[12] Initially we might think the topic of possible worlds has little to do with the way the Universe is composed of many u-fibers, but we shall be assimilating David Lewis's modal realism to Multiverse.

Fine-Tuning

The laws that scientists have discovered involve various constants, such as the charge of an electron and the fine-structure constant, that have to be fine-tuned if the conditions for life are to arise. Those of us who expect simple fundamental laws hope that we can avoid the inclusion of any such constants. To do so, we may hypothesize fundamental laws that have no such constants and so are not subject to fine-tuning. On this hypothesis there are derived laws, resulting perhaps from events at about the time of the Big Bang. These derived laws vary in different parts of the Universe, resulting in an immense variety of values for the constants. Hence in some parts the laws result in the wonders of chemistry and, moreover, chemistry suited to life. Naturalists are especially eager to avoid fine-tuned constants but even theists should expect simple fundamental laws. As far as I can see, any multiverse hypothesis coheres well with this, admittedly speculative, explanation of fine-tuning. Indeed it would suffice just to have the one u-fiber with many domains in it or many epochs. So when I come to compare multiverse hypotheses, I shall not take fine-tuning into consideration.

Apparent Indeterminacy

Cross Fiber Extension explains apparent indeterminacy. Instead of saying that our universe is indeterminate between states X and Y, we may suppose that our universe as of now is a bundle of u-fibers including x and y, with X occurring in x and Y in y. There are two points worth noting about this explanation of apparent indeterminacy. The first is that a person's mental states in the universes x and y should be compatible. Given that a certain degree of confusion is my normal psychological state—yours too, maybe— this condition is somewhat vague. The second is that not all conceivable truth-value gaps or gluts have to be treated as indeterminacy. Or, if they do, then they do not all admit of multiverse interpretations. For example, I have been persuaded by Graham Priest that the infamous liar statement ('What is hereby asserted is not true') is both true and false, but that is not because it is true in some u-fibers and false in others: it is true and false everywhere at all times.

I do not treat as a distinct application the many 'worlds' interpretation of quantum theory, suggested by Everett (1956), made more explicit by De Witt (1970), and championed by Deutsch (1998). For all these theories, in both their collapse and, more usual, no-collapse formulations, may be stated in two stages. First, quantum theory is interpreted in terms of (real or apparent) indeterminacy and then the multiverse explains the apparent indeterminacy.[13]

AGENCY AND FREEDOM

I now turn to some applications of Multiverse whose success supports that generic hypothesis, and which provides the context for the comparison of the different multiverse hypotheses. In this section I examine agency and related topics. First consider the problem of free will and deterministic laws. In Forrest (1985) I defended a version of compatibilism by the drastic means of saying that, given deterministic laws, we may freely affect the future by also freely affecting the past. This counterintuitive conclusion is the *reductio ad absurdum* of Peter van Inwagen's earlier argument against compatibilism (van Inwagen 1975), as I now grant. But there is a successor to my previous theory. Let us suppose that the human brain operates in a deterministic way in every u-fiber, but my current brain state is indeterminate between state X that leads, say, to the decision to vote for the Reddish Green party, and state Y, that leads to a decision to vote for the Greenish Red party. Then the decision I freely make increases the determinacy, by restricting the range of u-fibers between which there is indeterminacy.

This account is not a compatibilist one. For even if determinism holds in every u-fiber, it does not hold for the whole multiverse. The resulting theory of agency may be adapted to, but is independent of, a 'no-collapse' many-worlds interpretation of quantum theory. First I shall expound it in greater detail, and then I will consider no-collapse theories.

A Theory of Agency

As part of Properly Anthropocentric Metaphysics, I assume that agency is not explained in terms of laws of nature. This supports theism, by inviting us to turn the tables: agency explains laws (see Swinburne 1977, Chapter 2; Alston 1993), supporting theism. Now, this is compatible with some further nonnaturalistic *explanation* of agency, a topic beyond the scope of this essay can be given. Instead of explaining agency, I follow McCall (1994) in holding that Multiverse provides us with a more accurate *description* of agency than the commonsense alternative stated later.

I use the term 'act' not as a synonym for 'action' but to refer to whatever is the immediate effect in a case of agency causation, that by-or-in which the agent performs an action. The act is, I hold, the locus of freedom in that a necessary and sufficient condition for an *action* to be free is that the *act* is free. By an action, I mean the bodily movement associated with the act, together with attendant nerve processes, and some of what is brought about by-or-in the bodily movement.[14] *Action Itself* is the hypothesis that a human act is the Action Itself. I take this to be common sense, which, I concede, establishes a strong presumption in its favor. Because an action is describable in purely physical terms, Action Itself leads to what is perhaps the most serious difficulty with PAM, Smart's Dilemma: either an act is determined or else random, neither of which is acceptable to incompatibilist defenders of human freedom (Smart 1963). Moreover even if some version of compatibilism should be conceded by the friends of PAM, neither determinism nor randomness is compatible with the gravity of choice.

I shall now argue that we may undermine the support common sense provides for Action Itself, by noting that it is the natural enough result of merely thinking about *one* u-fiber.[15] Given Multiverse, an alternative description becomes at least as intuitive as the commonsense Action Itself. I call this alternative *Selection:* to act is to select some among the many u-fibers, making, as we say, a free informed choice. Indeed we may say that common sense is, as so often, confused because although it is common sense to identify the act with the action, the idea of an act is often expressed using the metaphor of a fork in the road (Robert Frost's choice of the 'road . . . less travelled by'). Because our universe splits when we act, with part of it being rejected, this metaphor is near the literal truth.

Selection requires Multiverse, while Action Itself is consistent with there being many u-fibers or just one. For example, suppose I perform an action, say, my voting for the Greenish Red party, when I might have voted for the Reddish Green party. According to Action Itself, some u-fibers are selected because they are ones in which I *voted* Greenish Red. According to Selection, some u-fibers are selected because they are the ones in which I *vote* Greenish Red. The difference is whether the selection of some but not other u-fibers is explained by my action or constitutes my act.

My reply to Smart's Dilemma is that because the past is somewhat indeterminate, what we do affects past chances. This is most easily seen if we suppose determinism within each u-fiber. Consider two times t_1 and t_2 both prior to my voting, with t_1 earlier than t_2. Suppose I make up my mind between t_1 and t_2 to vote Reddish Green. Then at t_1 it is true that the chance at t_1 of my voting Reddish Green is the proportion P of the then actual u-fibers in which some past state (at time t_0) and the deterministic laws jointly entail that I vote that way. The proportion P might be quite small. Because determinism holds within u-fibers, it is true at t_1 that the chance at t_2 of my voting Reddish Green is P. It is true at t_2, however, that the chance at *time t_1* of my voting Reddish Green is the proportion Q of the now actual (i.e., at time t_2) u-fibers in which the past state (at time t_0) and the deterministic laws jointly entail I vote that way. Q is 100 percent minus the small proportion in which I fail to execute my decision, for instance because I drop dead at the voting booth. So Q is greater than P. Again, because determinism holds within u-fibers it is true at t_2 that the chance at t_2 of my voting Reddish Green is Q.

This sensitivity of past chances to our acts shows that our acts are not random. Indeed, in hindsight whatever we choose was likely to be the case. It might be objected that because an act is not determined it may be called chance, even if, as I have submitted, the precise chance is affected by the subsequent act. This is, I think, a mischievous use of the word 'chance', and I could with equal justification say that for the naturalist it is just chance what laws there are. Again, theists could be accused of saying the existence of God is chance. But this is silly. The right word is not 'chance' but 'mystery'. The relevant difference is that to say something happened by chance implies that the numerical chance offers a partial explanation of the event. On my proposed account, however, the explanation is the other way round: our choices affect the chances.

I now turn to an obvious objection to Selection. When a sane person acts, the intention is that a certain action occurs, not that infinitely many u-fibers be rejected, or even that the remainder be selected. I agree, but this is a familiar feature of intentions, namely that what we intend to happen is not the most immediate effect of the mental process involved in acting. To take a well-known example, axon-firings in the motor nerves leading to the muscles in the hand can be caused by the events in the mind/brain that accompany the action of clenching the fist. Under normal circumstances, the intention is to clench the fist and the person might know nothing of nerves. In peculiar circumstances the same type of action could be used precisely so as to have produced those axon-firings. Likewise in peculiar circumstances we could use that or any other action to demonstrate our god-like power to exterminate some and preserve other universes-fibers—supposing for the moment that rejects are exterminated.

Agency and No-Collapse Theories

The 'many worlds', that is multiverse, interpretation of quantum theory, due to Everett, deWitt and Deutsch is a no-collapse hypothesis according to which if quantum theory states that a type of event E would occur with probability K after an observation, and it did occur, then an E occurred in proportion K of u-fibers. You and I, however, are only in (some of the) u-fibers in which an E occurred and in either none or a very small proportion of those in which an E does not occur. So there are other universes in which you and I do not now exist and it was observed that no E occurs. This result is said to hold in the case in which you and I were there before the observation. Thus even if there is no splitting of u-fibers, the observers, who are spread across many u-fibers, split along with the universe they inhabit. The upshot is that if we hear the next Geiger counter click in two seconds we have already split into observers who hear the Geiger counter one second later, two seconds later, and so on. This, like Lewis's token-reflexive theory of actuality is "refuted by an incredulous stare".[16] More precisely, it is counterintuitive, and has, I say, a semi-fatalist consequence that is additionally counterintuitive. This consequence is exhibited by the following Cassandra Argument.

Let us begin by considering Cassandra, in Aeschylus' *Agamemnon*, who foresees the circumstances of her own murder. Now these circumstances will depend on the choice she makes not to flee, but her own foreknowledge of the outcome robs that choice of its gravity. So even though not strictly fated, she may say, "Whatever will be will be".[17] Those who knew that the no-collapse theory was correct would be in the same position as Cassandra. They would know the outcome of their deliberations before the choice is made, namely that they would split into some who make one choice and some who make the others. I call this semi-fatalism because it is prospectively but not retrospectively fatalistic: the future outcome is independent of any choice you make but in hindsight things would have been different if you had chosen otherwise. I find semi-fatalism counterintuitive, and

I explicate this intuition by arguing that semi-fatalism undercuts the gravity of our choices. Maybe semi-fatalism would result in a welcome freedom from anxiety but, welcome or not, it is contrary to our intuitions about decisions. The Cassandra Argument is based on the premise that given the multiverse interpretation of physics, decisions result in a splitting of observers. This in turn follows from two premises: that the distinction between a free choice and a random event is invisible to physics; and that all random events are like quantum-theoretic observations. The second is supported by intellectual economy—why posit randomness in addition to the probabilities inherent in quantum theory? The first premise is more controversial and would be rebutted by causal determinism for those cases in which we are usually taken to make a free informed choice when faced with a dilemma. Here I assume that the gravity of our choices would be undercut by determinism, which establishes a presumption against determinism.

I draw the conclusion that multiverse interpretations of quantum theory need some surrogate for the collapse of the wave-packet: in some way or another we observers fail to split. Storrs McCall (1994) proposes that the branches of the Branching Universe 'fall off', that is, all rejects are exterminated. Thus the passage of time consists in the successive shrinkage of reality. So the stuff of the physical universe—the sum of ordinary particulars—can cease to exist but can never come into being. This way of 'collapsing the wave-packet' can be adapted to Separate Worlds or to Hyperspace.

A variant on this is to reduce the weight of what is rejected rather than completely annihilate it. This would result in an imperfect determinacy of the choice. Thus the result of my choice would be Reddish Green with just a smidgeon of Greenish Red. As a theory of agency this variant has little to commend it, but perhaps it coheres better with quantum theory. For simplicity of exposition I shall ignore it.

Rejection as Annihilation is not compatible with standard no-collapse theories. An alternative, which is compatible, is Rejection as Vanishing—namely that the rejected u-fibers still exist physically but no longer appear to any observer. If the observers are nonphysical egos, then this may be because these egos have withdrawn from the rejected u-fibers. If there are no such things as nonphysical egos, and a mind is a bundle of ways things appear, then we should just say that the rejected universes-fibers cease to appear.

Possible Worlds

It could be objected that whatever qualification is made to Multiverse to avoid its counterintuitive consequences, the result will be *ad hoc*, and, therefore, we should accept the unqualified, no-collapse, version. To reply to this charge of being *ad hoc*, I, assimilate the theory of many u-fibers to Lewis's theory of many possible worlds, which, has many metaphysical advantages provided, I say, we ignore his token-reflexive theory of actuality.[18] The problems with the no-collapse interpretation of quantum theory are some of the problems faced by Lewis's token-reflexive theory.[19] In both cases, I say, we need to distinguish between those physical worlds (u-fibers) that are actual from those that are merely possible, and we need to do so without resort to his token-reflexive account of actuality.

I anticipate two objections. The first is that we need to consider more than just possible spatiotemporal physical worlds, which are identified with u-fibers. For a theory of possibility requires possible histories, with possible acts. My reply to this objection is that we only require enough possible worlds to do modal semantics. It suffices to rely on a few straightforward constructions to obtain all the 'abstract' or, as Lewis would say, 'ersatz' worlds that correspond to maximal consistent sets of propositions. Our actual history is a sequence of sums of u-fibers, with later worlds parts of earlier ones. Possible histories are likewise sequences of sums of u-fibers, except that those that are merely possible include some u-fibers that have been rejected.

The second anticipated objection is that assimilating my speculation to a possible worlds theory with primitive actuality provides rather little support, given our lack of a good grasp of the actual/merely possible distinction. My response is that rejection provides us with a theory of actuality, which need not therefore be taken as primitive. The u-fibers that are actual are just those that have not been rejected.[20] On a dynamic theory of time what is actual at a given time gets less and less.[21] I concede that this reply requires the rejected u-fibers to leave traces of the futures they would have had. The most straightforward way of providing such traces is to deny that they terminate and to take the rejected to have vanished, that is, ceased to be objects of consciousness.

The conclusion we should draw is that the very same selection process is required to distinguish the actual world as required to qualify the no-collapse interpretation.

ONE PAST, MANY FUTURES?

As a friend of PAM, I am assuming a dynamic theory of time, in which past, (the specious) present, and future are of different intrinsic character. I shall be arguing for the indeterminacy of the past. This not only supports Multiverse, but provides an important negative point of comparison; namely that, when it comes to providing a dynamic theory of time, Branching has no advantage over other multiverse theories.

By a *standard* dynamic theory I mean one that accepts the intuitively appealing thesis, attributed to Aristotle, that propositions about past events (e.g., that there was a sea battle yesterday) are either true or false but not both, while propositions about future events (e.g., that there will be a sea battle tomorrow) are neither true nor false. I shall argue later for the indeterminacy of the past at the scale for which quantum theory applies. So I shall be retreating to a nonstandard theory that the present marks the boundary between the *macroscopic* determinate and the *macroscopic* indeterminate.

The debate between different dynamic theories is complicated in two ways, however. The first is that we might want to hold different theories for different categories. Thus a presentist might grant that there exist universals whose instances are all past, but deny that there exist any wholly past particulars. To avoid that complication, I concentrate on ordinary particulars, such as material objects and events, which make up the u-fibers. The second complication is that there are various compatibilities between hypotheses based on the supposition that some entities can exist at more than one time. For instance, if we characterize presentism as the thesis that whatever exists *simpliciter* exists *now*, then this is compatible with some things existing in the past or future as well as the present.[22] I shall, therefore consider two negatively characterized theses: Narrow Presentism, the theory that no particular exists simpliciter at any time except the present; and No Futurism, the theory that some particulars exist simpliciter in the past but none in the future. McCall's (1994) Falling Branches makes up a third theory in my classification of standard dynamic theories of time. Multiverse will provide us with a nonstandard variant on Falling Branches.

I begin by rejecting Narrow Presentism, My reason is that we judge what we are veridically aware of as present to be a particular that exists simpliciter, but the present we are aware of is the specious present, which is extended. Because the length of the specious present varies, even for the same human being, and presumably varies across species, we cannot identify the specious with the true present. So what exists simpliciter is not restricted to the present.[23]

Next consider the No Future and the Falling Branches theory. On the former, there is no future (yet); on the latter there are many futures and hence (the appearance of) an indeterminate future. On No Future, as time passes, new particulars comes into being, layer by layer. On Falling Branches, as time passes, there is less and less (of the sum of ordinary particulars) because there are many future branches but only the one past.[24] Falling Branches requires Branching. By contrast, No Future is compatible with there being a single u-fiber. Combined with the rejection of Multiverse, I call it One Fiber No Future. It coheres better with Action Itself than with Selection, because if we think of reality being added on layer by layer with the passage of time then agents could influence the next layer, but on One Fiber No Future there are no futures to choose between.

Both Separate Worlds and Hyperspace can mimic Falling Branches if the u-fibers are all rather similar in the past but differ considerably in the future. In that case, the passage of time requires that many u-fibers are 'rejected', either by annihilation or vanishing, just as in the selection theory of agency.

The important difference between One Fiber No Future, on the one hand, and the multiverse on the other, is that the former entails a precise present, the boundary between past and future; while the latter is consistent with a fuzzy present, that is an indeterminate past/future boundary.[25] This contrast is obscured a little by the way that Falling Branches offers some support for a precise boundary, although less than One Fiber No Future. For the idea of precisely one past and many futures has considerable intuitive appeal. That gives the initial advantage to Branching over other multiverse theories, but only on the assumption of a precise present.

Initially there is an advantage, then, to One Fiber No Future and to a lesser extent Branching as 'predicting' what we intuitively believe, namely that there is a true present, which is instantaneous and to be contrasted with the specious present. This intuition is attested to by that notoriously puzzled presentist St Augustine, who in *Confessions* Book XI, denies the reality of past and future, and hence worries about the vanishingly thin present.

I now argue for the fuzzy present. This is contrary both to Narrow Presentism and to One Fiber No Future, and so is an additional argument for the multiverse. It also nullifies the initial advantage of Branching as being a commonsense dynamic theory. The argument has the corollary that the past is not fully determinate.

The Case for a Fuzzy Present and a Somewhat Indeterminate Past

I assume the present is the boundary between the determinate macroscopic past and the indeterminate macroscopic future. Hence to argue for a fuzzy present it suffices to argue that at the quantum level there is some past indeterminacy. For in that case the imprecision of the boundary between the quantum-theoretic and the macroscopic results in a corresponding imprecision between the determinate macroscopic past and indeterminate macroscopic future. Consider the idealized experiment of the electron passing through two slits. When does it become determinate as to which slit it goes through? Unless we invoke backward causation or some mysterious action at a distance, it must still be indeterminate when one of the two slits could still be shut. But given a plausible setup for the experiment, it would then be too late for the electron to change its course to go through the other slit unless it exceeds the speed of light. The alternatives to past indeterminacy are all, then, radical theses: backward causation, action at a distance, or exceeding the speed of light. With respect to a suitable relativistic frame of reference, all three of these involve backward causation, which is counterintuitive.

An alternative route to the same conclusion is that the state of a quantum system consisting of a finite number of particles is specified by a function that does not contain enough information to specify both the positions and the momenta of individual particles—note in this connection the Heisenberg Uncertainty Principle. Among these states are ones of fairly determinate momenta and hence highly indeterminate positions. The manifest interpretation of this is that *at all times*, including the past, the states are indeterminate subject to the proviso that indeterminacy may be paraphrased, as I have, in terms of multiple universes. To be sure a hidden variable theory, such as David Bohme's (Bohme and Hiley 1993) or a carrier wave interpretation such as Henry Krips' (1990) would enable us to resist this conclusion, but at the cost of increased complexity.

PROBABILITIES

Another advantage of Multiverse is its capacity to explain objective chances (or physical probabilities) as proportions. This is significant because one of the many puzzling features of quantum theory is the way the quantum state specifies probability distributions, for part of the puzzle is that it is not clear just how the dots are to be filled in, but at the risk of circularity let us just say that the probabilities are for the results of idealized two-valued (i.e., Yes/No) observations.[26] An important case is the observation of whether there is a particle in a given region.[27] I interpret the probability of a Yes result as a relative probability of universes V and W, that is as Prob(V/W) where $V \subseteq W$. Here W is our universe as of a moment before the observation and V our universe in case we observe the 'Yes' result.

If W contains a finite number of u-fibers, then we can explain the probability by taking Prob(V/W) to be $\#V^*/\#W^*$—the ratio of the cardinalities of the sets V^* and W^* of the u-fibers in V and W respectively. This is not intended as a piece of conceptual analysis, for a curmudgeon might protest the assumption that all u-fibers are equally probable. It is, however, a satisfactory explanation of the objective character of the probability.

Unfortunately, even if Space-time (or better the underlying aether; see Forrest 2012b) is discrete, that is, has a finite number of simple parts per unit volume, $\#W^*$ is infinite if the past is infinite or if at least some of the u-fibers are spatially infinite, or if some of them have infinite futures.[28] For consider the number of ways a simple part x of Space-time can be related to its neighbors or have intrinsic properties. We anticipate that this number is at least 2 for all x. Hence we anticipate that the number of ways n mutually distant parts can be is roughly 2^n, and so there is an infinity of u-fibers in our universe, even if Space-time is discrete. Using standard numbers, it would seem, then, that $\#V^*/\#W^*$ is ∞/∞ and so undefined.

There are two ways of dealing with ∞/∞, taking limits and using nonstandard analysis. The former is more familiar, but readers who find it counterintuitive should employ the latter and refer to "On Multiverses and Infinite Numbers" by Gwiazda (Chapter 8 in this volume). For example, consider the sequence of reciprocals of natural numbers, 1, 1/2, 1/3, and so on. We may say its mean value is the limit of the mean of the first m members as m tends to infinity, namely 0. Fortunately neither method is strictly required—we need one or the other only to avoid abstruse mathematics. (For which see the footnotes.)

I propose two ways of characterizing proportions. Again I emphasize that this is not intended as conceptual analysis but merely as providing some objective grounds for the statement of probability—a truth-maker, if you like. I begin with a method applicable to Branching but which uses limits and then expound a method that, although applicable to all versions of Multiverse, is not plausible in the case of Separate Worlds. On my preferred version, no limits are required. (I note in passing that proposing a primitive quantity of existence (Vaidman 2014) to characterize probabilities, is an *ad hoc* addition to a hypothesis and so to be avoided.)

On Branching, probabilities are defined if at each branch point there are only two, or more generally finitely many, branches (see McCall 1994).[29] Then the probability of a u-fiber is that like that of a sequence of heads and tails in the toss of a fair coin. This is the most straightforward way of obtaining the probabilities. If there are infinitely many branch points, then the sequence of coin tosses is likewise infinite and the overall probabilities could be defined as infinite products, which are limits of finite products. Because no additional structure needs to be hypothesized, Branching has an advantage over other multiverse hypotheses, when it comes to probabilities.

The more general method is to rely upon hypervolume. This is best expounded by supposing there to be a flat finite dimensional Hyperspace. Given a choice of coordinates, there is a hypervolume, that is, a measure of the quantity of a suitable region of Hyperspace, and hyperballs of various radii. In that case we may characterize relative probabilities for universes V and W in terms of the hypervolume, hvol. The idea is that even though V and W are of infinite hypervolume, we may restrict attention to a very large but finite part of hyperspace, provided it makes negligible difference which large part we consider. So we may take $Prob(V/W)$ to be the limit of the ratio of $hvol(V \wedge Z)$ to $hvol(Z)$ as we consider larger and larger parts Z of W. The choice of coordinates does not affect the end result of taking the limit, and greater attention to the mathematics would show that we do not need to consider coordinates at all.[30]

There are several ways of adapting this to other hypotheses about hyperspace. If hyperspace is discrete and finite dimensional but not flat, we may define hypervolume by assuming all its atomic parts are of the same hypervolume. If hyperspace is continuous and not flat but has some other highly symmetrical shape such as that of de Sitter space, then we may still define hypervolume as the (up to a multiplicative constant) unique measure invariant under the symmetries.

An important generalization is that in which universe W fits into a hyperspace that is not the whole multiverse, perhaps because the multiverse is a much bigger hyperspace containing all physical possibilities. In that case we restrict attention to the smallest suitable (e.g., flat finite dimensional) hyperspace containing W.

We may also generalize this procedure to apply to the case of Separate Worlds by defining distance as dissimilarity and then hypothesizing that a suitable part of the multiverse, containing W, will either be discrete and finite-dimensional itself, or will have the abstract mathematical structure of a flat, or otherwise symmetric, hyperspace. This application to Separate Worlds does not, however, provide a satisfactory explanation of the objectivity of the probabilities. To be sure, it enables us not to treat them as primitive unanalyzable quantities, but it fails to explain why quantum theory concerns a genuine probability of an event occurring, rather than some quantity with the mathematical properties of probability. The difference between the Hyperspace and Separate Worlds in this regard is that hypervolume has a geometric structure such as flat hyperspace. Hence it has a natural measure of quantity, an answer to the question 'How much?' By contrast, a hypervolume-like measure on a structure that is isomorphic to a flat hyperspace is merely something with the same mathematical properties as a measure of quantity.

I anticipate the objection that hyperspace should itself be lumpy and bumpy much as the u-fibers that make it up are (on the common interpretation of General Relativity in terms of curved Space-time.) I reply that the contorted u-fibers may be packed into a flat, or otherwise symmetric, hyperspace, like folded rock strata in a cube of rock of side one kilometer.

In spite of my enthusiasm for Hyperspace I concede that Branching provides the superior account of probabilities in the continuous case. For it does not require the additional hypothesis of a certain kind of geometric structure such as flat hyperspace.

THE REGION COUNTERPART PROBLEM AND
THE REJECTION OF SEPARATE WORLDS

The way Multiverse explains probabilities provides us with an *initial* comparison: Branching beats Hyperspace, which beats Separate Worlds. To continue the comparison I examine the Region Counterpart Problem. Most of the applications of Multiverse require particulars in one universe-fiber to be identified or correlated with particular other u-fibers. Thus the explanation of indeterminacy and the interpretation of quantum theory both consider an object x, an electron, say, that is apparently indeterminate between having properties U and V. This apparent indeterminacy is explained by x having parts u and v where u has U and v has V. Now u and v are not identical, for they have different properties U and V respectively. So we should take u and v to be counterparts. Because some objects considered in quantum theory, such as photons and other bosons, are far from unique, to specify their counterparts we need to specify the counterparts for the region in which they occur.

Now the shape and size of a region is not sufficient to specify counterparts because even if Space-time is discrete, there are too many candidates with the correct shape and size. The identity of some but not other regions of Space-time in different u-fibers is constitutive of Branching. But what we are considering in the applications of Multiverse are regions in which different events occur, so these regions cannot be identical.

The problem, then, is the difficulty of ensuring there are counterparts for regions of Space-time in distinct u-fibers. I call this the Region Counterpart Problem. I show how it can be solved easily using Hyperspace or Branching and then argue that Separate Worlds can solve it only by an *ad hoc* resort to *quissness* (quasi-thisness).

First consider Hyperspace. The simplest way of assigning counterparts occurs if regions are sums of points, but we can adapt it to a no-point theory. If there are points, the counterpart v in y to a point u in x is just the nearest point in y to x. Then the counterpart to a whole region u in x is sum of the counterparts to the point parts of u.

If Branching is correct, the counterpart relation between regions x and y in u-fibers v and w (after branching) may be characterized in a natural way using the backward time translations B^t that send a region z back in time by t. Regions x and y are counterparts if, for all large enough t, $B^t(x) = B^t(y)$. That is, x and y coincide if we go back to before the branching. The result is independent of the choice of frame, for it may be restated thus:

Regions x in u-fiber v and y in u-fiber w are counterparts if there are translations S in v and T in w such that: (1) In any common part of v and w, S and T coincide; and (2) $S(x) = T(y)$.

To be sure, I have here assumed there are translations in the u-fibers, which requires them to be flat, but we could generalize, replacing straight lines by geodesics.

Separate Worlds does not permit an analogous solution to the Region Counterpart Problem. It might seem we could solve it by positing thisnesses for regions generally or maybe just for points. These would be entities such that no two regions in a given u-fiber share the same thisness but a given region shares its thisness with many others in different u-fibers. I reject this on the grounds that we are here dealing with many actual u-fibers, in which we and other objects are spread out. Whatever the merits of positing thisness instantiated once only in a given possible worlds, what is here being posited is not a thisness but a new kind of property, a *quissness*, that happens to have the curious feature of just one instance per u-fiber.

This leaves one final Separate Worlds attempt at solving the Region Counterpart Problem, namely supposing there is just one way of correlating regions in two similar u-fibers that maximizes the resemblance of the objects and events in correlated regions. Suppose we can correlate u-fibers v and w so the occupants of regions resemble each other. Now consider the u-fiber v′ that exactly resembles v but everything is displaced by a light year. Then the attempt to correlate v′ with w would be a complete failure. The u-fibers in which the objects and events are repeated (almost exactly) at regular intervals prevent correlation, but the applications of Multiverse can cope with a few exceptions to the existence of counterparts to regions.

I have two objections to this solution, which complete this argument against Separate Worlds. The first is that quantum theory should hold even in a universe that is a quantum vacuum with probabilities for various events but no macroscopic objects or events that would serve to correlate the u-fibers. The second objection is that even though we do not yet have a fundamental 'theory of everything', quantum theory is, as far as we know, the correct framework for such a theory. Now a derived, that is nonfundamental, theory might be stated in terms of probabilities that are in turn characterized in terms of a counterpart relation that depends on complex objects and events. A fundamental theory requires, however, a simpler characterization of counterparts.

I conclude that we should definitely reject Separate Worlds for the u-fibers in the actual world, if quantum theory is fundamental, and so we require objective probabilities and counterparts in the fundamental theory.

More tentatively, I think we should reject Separate Worlds even if quantum theory is not fundamental and there is some underlying theory that does not imply that there are objective probabilities. In that case we, the friends of PAM, still need a multiverse for Selection as a theory of agency, and naturalists may still need the multiverse to deal with fine-tuning. Assuming, for one reason or another, then, that we retain Multiverse, a familiar criticism of David Lewis's modal realism can be used to provide an additional argument against Separate Worlds, even if we reject Lewis's token-reflexive account of actuality. This is Saul Kripke's Humphrey Objection. Counterfactuals about Humphrey winning the election are intuitively about Humphrey, not about a mere possible counterpart. I invite readers to agree that this shows that actual Humphrey-parts in various u-fibers must be related in such a way that that their sum has a unity, much as Lewis held that the various stages of actual Humphrey are united into a four-dimensional whole. The requirement here is not that the name 'Humphrey' refers to the unity, merely that there is the right sort of unity to count as one person. This can be achieved using Branching or Hyperspace but requires the *ad hoc* positing of quissness in the case of Separate Worlds.

CAN BRANCHING AND HYPERSPACE PROVIDE ENOUGH POSSIBLE WORLDS?

In the previous section, I assimilated the many u-fibers to the many worlds of modal realism. That is unproblematic on Separate Worlds, a position that I have, however, argued against, because of the counterpart problem for regions, and the Humphrey Objection. This leaves Branching, Hyperspace, and various mixtures such as Forest (i.e., many separate branching structures,) and Many (separate) Hyperspaces. The reason why those compromises with Separate Worlds are not totally refuted is that we only need to postulate counterparts of regions for similar worlds to ours, not all of them.

There is, however, an obvious reason for preferring Branching or Hyperspace to mixed hypotheses such as Forest or Many Hyperspaces. For these are less elegant and hence less satisfactory than the pure alternatives. The only reason we might have to accept them would be if a single hyperspace or tree is not large enough to contain all possible ones.

Is one hyperspace or the one tree big enough, then? Consider a hyperspace first. Any precise quantity of dissimilarity can be taken as a distance-function ('metric'). It is not plausible that there is a unique correct quantity of dissimilarity but different metrics can give the same topological structure. For there to be a single hyperspace, this topological space has to be connected. This requires that we pass by small degrees from one possible u-fiber, v with Space-time S to another v', with Space-time S'. To do this we consider a third u-fiber z that contains exact copies w and w' of v and v' attached by a thin thread. By small degrees we pass from v to u-fibers in which w' is whittled away until there is nothing left and then the thread is shortened by degrees until we arrive at u. Similarly we can connected v to v' by whittling away at w. Such a whittling away of v (or v') could be achieved by introducing countably many holes in v and expanding them until nothing is left. (We believe a priori that actual universes do not have holes but these are merely possible ones.)

Now there is no reason to believe that the topological space of all u-fibers is locally compact, and even if it were, there does not seem to be the required geometric structure (more precisely, a group of symmetries that acts transitively on the u-fibers). That was not, however, a requirement for the whole hyperspace, but merely for that part of the hyperspace that contains the u-fibers needed in the interpretation of quantum theory.

Branching, however, is subject to a serious problem of fitting in all the possible worlds. For it is plausible that in some possible world, maybe even the actual one, there is an objectively infinite past, by which I mean there is an infinite past and this cannot be redescribed as one with a finite past, which would be a different world. It is likewise plausible that there be worlds with objectively infinite pasts that are at no time shared, perhaps because of a difference in the fundamental laws of nature or the number of dimensions, but maybe merely because of differences in details. To fit such worlds into a single tree we require some moment of time an infinite time ago.[31] Whether this is even possible is doubtful, but we are considering the actual physical Universe, not something merely possible, and we know a priori—or at least have a firm intuition—that there is no actual moment an infinite time ago.[32]

This will lead to the Uncomfortable Trilemma. I call it that because the horns are not sharp enough to be lethal. Either prefer Hyperspace to Branching in spite of the latter's other advantages; or adopt the unsatisfactorily mixed hypothesis of Forest; or restrict the range of the metaphysically possible in violation of Hume's Razor, the principle of not multiplying metaphysical necessities (see Forrest 2009).

THE PRIMITIVE FIBRATION PROBLEM FOR HYPERSPACE

There is an objective division of hyperspace into u-fibers. If this is an unanalyzed addition to the geometry of hyperspace, then that is a disadvantage and the first horn of the Uncomfortable Trilemma is somewhat sharper. For by comparison the distinction between the u-fibers on Branching is determined by the topology of the multiverse.[33]

It is not difficult to speculate about the physics of the fibration: rest mass cannot be transferred across u-fibers because the Higg's field—or maybe something else, the quiggs field—varies across u-fibers.[34] Hence we may characterize the u-fibers using barriers to rest mass transfer. Clearly such arrant speculation is little better than the retreat to primitives.

COMPARISONS

I have already rejected Separate Worlds. The comparison of Branching (including Lattice) with Hyperspace is not so clear. Branching has some advantages, such as providing a straightforward theory of probabilities and avoiding a primitive fibration. On the other hand, the Humphrey Objection, which favors Hyperspace over Separate Worlds, offers some support for Hyperspace over Branching. For on the former, the parts of actual Humphrey in different u-fibers are joined in a fashion analogous to person-stages. Whereas on Branching, these parts are no more closely related, it would seem, than products of fission, and so are analogous to identical twins, and hence not the same person.[35] This leaves us with the Uncomfortable Trilemma stated earlier:

> Either prefer Hyperspace to Branching in spite of the latter's other advantages; or adopt the unsatisfactorily mixed hypothesis of Forest; or restrict the range of the metaphysically possible in violation of Hume's Razor.

I now argue that theists should prefer Hyperspace to Branching.

First suppose you assess the 'other advantages' of Branching as negligible. Then clearly Hyperspace should win. This, I say, leaves the case in which the deciding factor is the presumption in favor of a very broad range of possible worlds. This presumption is stronger for theists who hold the Selection theory of agency than for atheists. For any restriction to the possible worlds is a restriction on divine power. But the presumption, although weaker, is problematic for any atheist friends of PAM, who cannot rely upon a primal act of God to restrict the range of the actual to the physically possible. The further restriction of what is actual could indeed be due to finite agents such as us, but not the restriction to the most fundamental (and I say not in need of fine-tuning) laws. Once the atheist acknowledges this problem, there is no further argument against Branching based on the difficulty of fitting all the trees into one tree. So, unless Hyperspace is otherwise more attractive, the atheist may well resort to Branching.

CREATION

I have argued that theists should prefer Hyperspace to its rivals. I now consider what difference the choice of multiverse theory makes to theism. I begin by considering the act of creation. A general theory of agency should have applications to divine agency, including an initial act of creation. For to posit some *sui generis* divine act in addition to the agency exemplified by human beings is an obvious violation of Occam's razor, although surprisingly popular.[36]

Action Itself, which I have argued against, leads to an account of creation congenial to those pure classical theists, such as Leftow (2012), who hold the difficult thesis of divine simplicity and who, in addition, insist that any nondivine necessary beings depend for their existence on the divine will. For they should not believe there is a preexisting nondivine realm of possibilities between which God chooses, nor, to preserve divine simplicity, should they think of these possibilities as ideas in the mind of God or even the intentional objects of a multitude of divine imaginings.

Because pure classical theists must reject Selection, I buttress my case for it by noting its theological advantage. For, I submit, the goodness of creation is a consequence of the way *any* agent will be guided by the (in general, real or apparent, but real in the divine case) value of the options presented to the agent (Forrest 2012b). Now I accept that latter thesis because it is more economic not to posit an initial divine moral character as an essential attribute.[37] I also note that it coheres well with the theodicy that (initially at least) the divine goodness is not concerned with present individuals and God creates as a wise utilitarian. (In the divine case, God can presumably compensate individuals for suffering in an afterlife, so the lack of love exhibited by divine utilitarianism is not manifested by sacrificing some for others but by sacrificing the present for the future.)[38] So I am pleased that many universe hypotheses cohere poorly with Action Itself and the associated theory of creation.

The theodicy of (initial) divine utilitarianism is, I now believe, redundant given a selection theory of creation. But because this redundancy is based on Selection it does not weaken the theological case against Action Itself. Suppose God initially was a morally perfect being and so of a totally loving and compassionate disposition. If God creates *ex nihilo*, such a God would not, we judge, make it likely that there will be creatures who suffer immensely even though they will receive ample compensation. So the sort of 'soul-destroying' suffering that many do experience would have to be considered the unintended consequence of some, probably Satanic, creaturely choice. While not condemning that theodicy, I am aware that it is not popular among intellectuals. I note, however, that a loving God will act differently if creation is selection. For instead of bringing into existence a universe prone to 'soul-destroying' suffering, God would be faced with the choice of which universes to damn into nonexistence. Love is, however, loath to destroy. An alternative way of reaching much the same conclusion is based on the premise that the default position differs for destruction and creation. Unless something is clearly worthy of continued existence it should not be created, but unless something is clearly not worthy of existence it should not be destroyed, but rather redeemed.

I propose, therefore, a selection theory of creation. God is aware of the many possible universes and selects some of them, leaving the others as mere might-have-beens. God, like any other agent, subsequently acts by restricting the range of universes, again leaving the remainder as might-have-beens. We may think of God as a master sculptor carving one or probably many rough figures out of a vast and variegated block of marble. We, the apprentices, then labor together on just one of these figures, with the master sculptor helping only if it becomes necessary.

On this selection theory of creation God has no supernatural power to bring things into existence *ex nihilo*. God acts like any other agent by selecting some out of a range of possible universes, leaving the rest as might-have-beens. It is not so much matter of God saying 'Let there be light!' but as God either annihilating or else casting the other universes into an eternal darkness wherein there is no consciousness. If that sounds like damnation, so be it.

The overall well-being of creatures will be ensured if God 'damns' those damnable universes that, like ours, are miserable for many, but which lack any compensating afterlife. So a happy corollary is that in our universe there will be a compensating afterlife.[39] Now this is not to deny that God is morally responsible for evil. It is rather to say that God was in a serious moral dilemma when deciding whether or not to 'damn', that is either annihilate or cause to vanish, our universe. It was not clearly unworthy of existence but still contained much evil. I consider the divine moral dilemma to be an acceptable theodicy unless we both assume that God cares for individuals, and recoil from the idea of God in a dilemma. There is, however, a successor problem, that of As If Recurrence. I shall state it and some other problems and then argue that they can be solved by appeal to Hyperspace, which already seems the best many universes theory.

The problem of As If Recurrence occurs if God has the power to create many universes that subsequently come to differ as a result of creaturely freedom or randomness. Assuming that more on-balance-good universes are better than fewer, the only candidate for a best of all possible acts of creation will, the argument goes, be to bring into existence a proper class of universes similar to our own. That generates a dilemma. One horn is that there is no best possible act of creation because God cannot create a proper class of universes, in which case it is hard to see how God chooses what to create (see Rowe 2004). The other horn is that God creates 'class many' instances of a given type of universe in which case, even given variations due to randomness and free choice we have something like Nietzchean eternal recurrence, except not in the same universe. Every situation in minutest detail occurs infinitely often. This is compatible with theism but contrary to the Abrahamic tradition, which holds that our relationship with God confers meaning on our lives as significant.

Even if we avoid this problem somehow, there is a further question concerning what God decides to actualize, if, for scientific or other metaphysical reasons, we posit an actual world that is larger than a single universe. How are the many universes unified into a single thing, so that we can say that God created it in a single act? The similarities between universes might be sufficient to unite them into a system, and hence a single object of consciousness, but any such unification would seem to imply there was a *set* of all the universes, contrary to the assumption that they form a *proper class*.

This problem should not be dismissed because it derives from the Kantian account of the unity of the conscious mind as due to and explained by the unity of the content of consciousness. For this Kantian account of the 'unity of apperception' is not supported merely by some empirical claim about the limitations of human brains. Our brains are processing information in many parallel streams, but only one of these streams is my consciousness at any one time. I suggest that an action starts with just such a stream of consciousness, so that in choosing how to act we choose one of several streams of consciousness. Even if I am wrong, the unification of consciousness is plausibly not to be treated as a physical limitation. Hence the only sort of anthropomorphism involved is that which seeks to explain the infinite difference between the divine and the human in terms of the local character of human embodiment, taking the correlation between the mental and the physical as a matter of metaphysical necessity.

One drastic solution to this problem is to posit a distinct divine mind for each possible world, with the power to create or not create that world.[40] Such polytheism is not compatible with the mildly heterodox Christian position that God is a Holy Infinitary, for it requires there to be a proper class of divine agents, which could no more be unified into a single God than the proper class of universes can be united into a single object of divine consciousness. In any case such polytheism does nothing to solve the problem of As If Recurrence.

There are two problems, then, that arise if God creates too much: As If Recurrence and Divine Disunity, with only the latter soluble by means of an infinity of divine agents. My solution to As If Recurrence is to exploit the way our universe contains many u-fibers, across which we are spread out. I submit that a person can be spread out so much as to include as parts what we might have thought of as counterparts in similar situations.

The problem of Divine Disunity is solved, I say, by means of the Hyperspace hypothesis—there is a topologically connected space of universes, providing the unity. It is a matter of further speculation if a version of the Identity of Indiscernibles holds or if there could be exact replicas of a given universes. If there could be, then they all occupy the same 'point' of the topological space of possible universes. And, I would suggest, not even God can distinguish them by direct awareness, although God might know that there are, say, a proper class of them. Hence they do not threaten the divine unity.

I argued previously that theists should prefer Hyperspace to Branching. The considerations of divine unity just discussed show that strict monotheists should definitely prefer Hyperspace to either of the mixed hypotheses Forest or Many Hyperspaces, which seems to require a god for each hyperspace. They should do so even if not already persuaded of the inferiority of mixed hypotheses.

While on the topic of creation, I note a serious theological problem with the Leibnizian hypothesis of preexisting possible universes waiting for actualization. For this implies either predestination or Molinism. In both cases, the whole universe that is created would seem to be a spatiotemporal entity with past, present, and future included, leaving no room after creation for any human decision to make a difference. Predestination is contrary to the intuition of the gravity of human decisions and, I would claim, also contrary to human freedom. Molinism is congenial in that God decides which universes to actualize as a result of knowing what creatures would freely do in those universes. As a result, our choices retain their intuitive gravity: we are acting out a script that we would choose if we were not just acting out a script! Molinism, however, is problematic, largely because if we inhabit a universe whose whole history is chosen by God, we lack any freedom to exercise and hence there is nothing to ground the counterfactual about what we would have chosen.[41]

A further reason for adopting the selection theory of creation rather than the *ex nihilo* creation of the whole universe, past, present, and future, is that selection solves this Leibnizian problem. God does not bring into existence a universe complete with its future but rather 'damns' some but not other u-fibers, leaving reality indeterminate between many ways we creature can further choose between.

There are other solutions, but they are somewhat restrictive of the sort of universe that God could create for the sake of free creatures. If we hold the creation ex nihilo orthodoxy, we do not have to think of God bringing into existence a universe with past, present and future all equally real. Instead, God could bring into existence an initial segment of a universe, that is, everything up to a given time (or everything before a given time). In the extreme case, this could be an initial event such as the Big Bang. But what it excludes is a universe that has always existed and has always contained some creatures capable of free action. We should be reluctant to treat this as a possible universe that God cannot create.

I now draw together these speculations. God creates by either annihilating much of the initial Hyperspace or causing it to vanish, becoming the land of zombies. Now it is widely held by theists that all nondivine particulars depend for their existence on God. But the initial Hyperspace is there prior to the first divine act, so how does it depend on God? To avoid any *ad hoc* initial creation of the Hyperspace *ex nihilo*, we are driven to the conclusion that the Hyperspace is God, or at least it is the divine body. So the divine acts are acts of self-formation, Creatures have for their bodies parts of the divine body and the free creaturely acts are acts of sharing in this self-formation. I go further and tie in the 'damnation' of the no longer actual. There is, I speculate, proprioceptive awareness of what is actual, and, I further speculate, minds are constituted by the proprioceptive awareness of bodies. God is therefore initially constituted by the awareness of the Hyperspace and so there is no divine mind distinct from the divine body. Hence I arrive at pantheism, but not a pantheism in which God is identified with our universe. Instead, God is identified with the actual Universe. Initially this actual Universe is an 'apeiron'—the infinite whole that is indefinite between all actualities.

CONCLUSIONS

I have been comparing different multiverse theories in the context of PAM. Separate Worlds results in problems assigning the probabilities that quantum theory deals with. Quantum theory itself is often taken to be a no-collapse theory, but I argue from the gravity of our choices that there must be some selection of universes. This coheres well with the Selection theory of agency, which I prefer to the commonsense Action Itself theory precisely because the latter fails to cohere well with contemporary science. I have argued against Separate Worlds and that theists should prefer Hyperspace to Branching. Hyperspace, which coheres well with monotheism, may well have other theological applications (see Hudson 2006, Chapters 7 and 8). The conclusion reached, a selection theory of agency combined with Hyperspace, provides a satisfactory, although heterodox, account of divine creation as the selection of some and not other universes. It leads to the further speculation that we should identify God with the actual Universe, which is initially the whole Universe, pregnant with all possibility.[42]

NOTES

1. Where the details of some science lack any prospect of empirical verification or falsification in the near future, I consider them to be in the science/metaphysics overlap. String Theory and Loop Quantum Gravity are examples of such mathematical metaphysics.
2. A topological manifold is defined as locally (topologically) equivalent to an open ball, where I use the term 'ball' generically to include discs in two-dimensional

manifolds and line intervals in one dimension, as well as hyperballs in higher dimensions. The sphere, cylinder, and torus are three familiar two-dimensional topological manifolds, but in higher dimensions there is an enormous variety. One well-known speculation is that ours is a spatially finite but unbounded universe, a hypersphere, that contracts to something rather small and then expands again. In that case the smallest size is considered the beginning of the Big Bang. The contraction and expansion are not topological notions, so topologically this is a four-dimensional analog of the cylinder. If space-time is discrete, these definitions have to be adjusted. See, for instance, Forrest (2012b).

3. Near the branch point, the space-time would not be equivalent to an open ball. In one dimension, for instance, branching occurs at a Y-shaped region, whereas the one-dimensional ball is just a line segment, that is, an I-shaped region. The Y-shaped region and the I-shaped region are not topologically equivalent.

4. A Y-shaped region is the sum of two overlapping I-shaped regions.

5. A further technicality will be ignored for the sake of exposition. Maybe there is no maximal division, but a succession of ever finer ones. In that case universes are constructs obtained using maximal filters of non-arbitrary divisions. (A filter with respect to an ordering ≤ is a subset X of the ordered items such that (1) if y∈F and y ≤ z then z∈F and (2) if y∈F and z∈F then there is some x∈F such that x ≤ y and x ≤ z. In this case the items are divisions into disjoint parts and the ordering u ≤ v is u's being finer than v.)

6. Lewis (1986, 202) attributes the distinction to Mark Johnson.

7. I use the phrase 'as of now' because the u-fibers are themselves temporally extended, and I do not want to imply that I am considering just a time-slice of a universe.

8. The phrase 'the universe as of now' is vague in two ways. First, my universe as of now and yours may differ slightly. Second, because there is no precise boundary between the particles that are and those that are not part of me, and because different particles might belong to different, but overlapping universes, even my universe as of now might be vague.

9. Events in distinct u-fibers might nonetheless be considered simultaneous because at one stage they were both present. But there is no way of correlating their locations, if by that we mean spatial locations.

10. If, however, the respects of variation V form a compact connected dimensional manifold of dimension n, there might well be a curve dense in V. Then we can take M = 1 even though n is very high. I doubt, however, that V is compact.

11. Somewhat more realistic would be a stack of de Sitter and anti-de Sitter spaces each of four dimensions, fitting together into the five-dimensional analog of Minkowski space, with a distinguished central light cone that does not belong to either a de Sitter or anti de-Sitter space.

12. In Lewis's (1986) modal realism, no possible world could be the sum of unrelated u-fibers, but Phil Bricker (2001) has shown us how to modify modal realism to avoid this consequence.

13. A quantum state for several particles specifies, among other things, a 'probability' distribution over the configurations of the particles. The 'probabilities' may be interpreted as measures of indeterminacy. That is the easy bit. The joint probability distribution for position and momentum can take negative values, which is alarming. And I continue to be puzzled by spin. Hopefully these puzzles can be solved by resort to the quantum vacuum. See Forrest (1988, 1997, 1999) for some attempts.

14. This definition of an action is consistent with, but weaker than, Ginet's. For he takes having an 'actish quality' as required for an action. He uses 'by'

much as I use 'by-or-in' (Ginet 1990, Chapter 1). His requirement that an action have an 'actish quality' derives from his not distinguishing acts from actions, and hence his implicit adherence to Action Itself. I submit that 'having an actish quality' could be replaced as a necessary condition for an action by: 'either having an act as an initial segment or being what immediately occurs by-or-in an act'.

15. In addition, we can rebut Action Itself using the premise that the Principle of Alternative Possibilities is required for human freedom. For Action Itself leads to Frankfurt counterexamples to that principle (Frankfurt 1969). See also Derek Baker's recent defense of a 'Platonic model' for agency (Baker 2012). This coheres better with Selection than with Action Itself.

16. The token-reflexive theory of actuality is the thesis that the world that is actual is just the world we inhabit, and other worlds are actual for their inhabitants.

17. This has been noted as a problem for divine foreknowledge by Richard La Croix (1976). It would seem to undermine the gravity even of God's own decisions.

18. As described in Lewis (1986, Chapter 1). I used to think that uninstantiated universals could replace possible worlds, but it is hard to give an account of the structure of universals without assuming they have real instances. So those not instantiated in actual worlds are instantiated in merely possible ones. For a recent attempt at a theory of their structure, see Forrest (2006b). This is still, I think, too complicated.

19. This is not the place to rehearse these problems. For their exposition and, I say, inadequate replies, see Lewis (1986, Chapter 2).

20. Compare Wayne Davis's (1991) theory in which consciousness shifts from world to world. My proposal has the advantage of an initial indeterminate state prior to all choices.

21. In this way I offer an analysis of Michael Tooley's (1997) *actuality at a time,* which he takes as primitive in his otherwise excellent account of how we can have a dynamic theory of time without irreducible tenses.

22. I assume there is a primary sense of 'existence', which I call existence *simpliciter.* Then we have derived tensed senses: existence now, past existence, future existence, and, finally omnitemporal existence, namely, the disjunction of those tensed senses.

23. I am grateful to Peter Farleigh for drawing my attention to the significance of the specious present.

24. Bearing in mind that states of affairs and universals might accumulate as the ordinary particulars cease to exist.

25. This difference needs to be qualified by the possibility on No Future of holes in the past that get filled in later, or even a block that grows at both ends. Nonetheless, at each location in space-time it is all or nothing—either complete determinate reality or nothing at all. Unrelated Universes and Hyperspace differ from No Future in that there could be two universes b and c that differ only, say, because in universe c the spins of a pair of electrons are the opposites for the corresponding pair in universe b. This could result in past indeterminacy without there being a hole in the past yet to be filled in.

26. The circularity occurs if we interpret the idealization to be that which is predicted by the theory absent disturbances.

27. In the relativistic context the probability distribution is replaced by a vector whose components are the probability density and the three components of a probability current. Note also that the true observable is the number of particles in a given region in addition to those expected from the quantum vacuum. This can be negative if the number is below par. See Forrest (1999).

28. This is not a strict entailment because it is conceivable that at each moment only finitely many future possibilities leads to our universe lasting for ever. But that is implausible.

29. Suppose at every instant in a subset T* of the set T of all instants, the Universe divides in two. Then the set of all branches may be given the topological structure of the product $\Pi_t S_t$ of sets S_t of cardinality two, one for each instant t in T*. See (Woo 2013). There is a probability measure derived from giving the two members of each set S_t equal probability. It does not require limits for its definition because it is the unique measure that for any t in T* is invariant under the transposition of the two members of S_t McCall (1994) considers a special case, but we may consider any subset including the case in which T* = T.

30. To avoid limits, consider the *base space* (i.e. the topological space whose 'points' are the u-fibers). Assume this has some additional, geometric, structure, with a group of symmetries. In the example with the u-fibers taken to be cylinders, the base space is the set of radii. The symmetries consist of multiplication of the radii by a positive real number. This is the multiplicative group of the positive real numbers. If the group acts transitively (i.e., any u-fiber is mapped to any u-fiber by some symmetry, as in the cylinder example) and the group is suitable (i.e., locally compact, as is, for instance, the multiplicative group of the positive real numbers) then there is at least one measure on the set of u-fibers that is invariant under the action of the symmetries. (This is derived from the Haar measure.) Moreover any two such measures differ only by a constant of proportionality. We may characterize relative probabilities for sets of u-fibers X and Y of finite measure thus: Prob(X/Y) = $\mu(X \cap Y)/\mu(Y)$, where μ is any invariant measure. This characterization succeeds only when $\mu(Y)$ is finite (or $\mu(X \cap Y) = 0$). The finiteness of $\mu(Y)$ is plausible if an unbounded variation across a set of u-fibers would be noticeable at the macroscopic level, in which case our universe corresponds to a bounded set of u-fibers in the base-space. For the multiplicative group of the positive real numbers, the ratio of probabilities is the ratio of the logarithms of the standard measures of the sets.

31. To fit copies v and w of two trees into a single tree, the copies have to have a common 'tree trunk' preceding them. Consider a moment x in that 'tree trunk'. If it were a finite time ago, then since it precedes both v and w, they have a finite past, contrary to the assumption. Hence x is an infinite time ago.

32. I am not denying that it makes sense to talk of a moment an infinite time ago. (See, e.g., Chapter 8 in this volume.) What I deny is that there actually is such a moment. My confidence derives from the Tristram Shandy paradox and other supertasks that could have been completed from a starting point an infinite time ago. See, e.g., Craig and Smith (1993).

33. In the discrete case the (generalized) topology can be used to characterize the fibration in the hyperspace case, by saying that within fibers there are direct space-like adjacencies, whereas between fibers there are only direct time-like adjacencies.

34. I suspect this would result in a fibration that was relative to a Relativistic frame of reference, a variant noted in the second section of this chapter.

35. Another problem with Branching occurs if we assume both that, in each u-fiber, space-time (or the stuff that fills space-time) is continuous, not discrete, and that it is curved, not flat. Neither assumption is very well-supported but they amount to something of an orthodoxy. Together they raise the problem of how to characterize the differentiable manifold structure without positing an excessively complicated ontology. One way round this problem, if it is

a problem, is to assume the curved u-fibers are parts of a flat (or otherwise highly symmetric) hyperspace. (See Forrest 2012b.) This is potentially an advantage for Hyperspace over Branching (including Lattice). I suggest that this advantage cancels out the disadvantage that, to give an account of the probabilities, we require just such a geometry for hyperspace, unless space-time is discrete.

36. People often say they believe there is something divine but reject as naive the belief in a personal God. Either they mean by a personal God the Abrahamic God who loves individuals as such, in which case they are rejecting Abrahamism, no doubt because of the prevalence of evil around us, or they mean to reject agency as an ultimate explanation. In the latter case it is hard to see why we should believe in anything divine at all because the ultimate explanation will be in terms of laws of nature.

37. Pure classical theists would deny the need to posit an extra essential attribute of divine goodness, on the grounds that all the divine essential attributes are identical to God. To this I reply as follows. Although omnipotence, omniscience and rationality form a package that we may plausibly think of as constituting a single divine property, a positive moral character seems like something additional, and hence puts intellectual strain on the thesis of divine simplicity itself. Consequently the admittedly attractive thesis of (initial) divine simplicity is best served by a theory in which God makes a rational choice between independently existing possibilities, without being influenced by a property of goodness in addition to rationality.

38. See Forrest (2006a) for an account of why God might subsequently choose to be loving.

39. The problem of establishing that those resurrected in an afterlife are the same as those who have died (the 'replica objection') may be solved by requiring that the created universe itself contains laws of nature that will result in the resurrection of the dead. In that case, both this life and the next are brought about by-or-in the act of creating.

40. There are also some interpretations of Mormonism that are relevant here that I shall not expand on, except to say that I think the Church of Jesus Christ of the Latter Day Saints is much stronger in its metaphysics than its history.

41. There is an extensive literature on Middle Knowledge. See, e.g., Perszyk (2011).

42. I am grateful to James Franklin for a discussion of the mathematical structure of the class of all possible universe. I am also most grateful to Jeremy Gwiazda and Klaas Kraay for their helpful comments.

REFERENCES

Alston, W. P. 1993. "Divine Action, Human Freedom, and the Laws of Nature." In *Quantum Cosmology and the Laws of Nature: Scientific Perspectives on Divine Action*, edited by R.J. Russell, N. Murphy, and C.J. Ishim, 185–206. Vatican City: Vatican Observatory.

Baker, D. 2012. "Knowing Yourself—And Giving Up On Your Own Agency in the Process." *Australasian Journal of Philosophy* 90: 641–56.

Bohme, D., and B. Hiley. 1993. *The Undivided Universe: An Ontological Interpretation of Quantum Theory*. New York: Routledge.

Bricker, P. 2001. "Island Universes and the Analysis of Modality." In *Reality and Humean Supervenience: Essays on the Philosophy of David Lewis*, edited by G. Preyer and F. Siebelt, 27–55. Lanham, MD: Rowman and Littlefield.

Craig, W. L., and Q. Smith. 1993. *Theism, Atheism, and Big Bang Cosmology.* Oxford: Clarendon Press.

Davis, W. 1991. "The World-Shift Theory of Free Choice." *Australasian Journal of Philosophy* 69: 206–11.

Deutsch, D. 1998. *The Fabric of Reality: The Science of Parallel Universes and Its Implications,* Harmondsworth, UK: Penguin.

De Witt, B. S. 1970. "Quantum Mechanics and Reality: Could the Solution to the Dilemma of Indeterminism Be a Universe in Which All Possible Outcomes of an Experiment Actually Occur?," *Physics Today* 23: 30–40.

Everett, H. 1956. *Theory of the Universal Wavefunction.* PhD thesis, Princeton University.

Forrest, P. 1985. "Backward Causation in Defence of Free Will." *Mind* 94: 210–17.

Forrest, P. 1988. *Quantum Metaphysics.* Oxford: Blackwell.

Forrest, P. 1997. "Common Sense and a 'Wigner–Dirac' Approach to Quantum Mechanics." *The Monist* 80: 131–59.

Forrest, P. 1999. "In Defence of the Phase Space Picture." *Synthese* 119: 299–311.

Forrest, P. 2006a. *Developmental Theism: From Pure Will to Unbounded Love.* Oxford: Oxford University Press.

Forrest, P. 2006b. "The Operator Theory of Instantiation." *Australasian Journal of Philosophy* 84: 213–28.

Forrest, P. 2007. "The Tree of Life: Agency and Immortality in a Metaphysics Inspired by Quantum Theory." In *Persons: Human and Divine*, edited by P. van Inwagen and D. Zimmerman, 254–53. Oxford: Oxford University Press.

Forrest, P. 2009. "Razor Arguments." In *The Routledge Companion to Metaphysics*, edited by R. Le Poidevin, S. Peters, M. Andrew, and R. Cameron, 246–55. New York: Routledge.

Forrest, P. 2012b. *The Necessary Structure of the All-Pervading Aether.* Frankfurt: Ontos Verlag.

Frankfurt, H. G. 1969. "Alternate Possibilities and Moral Responsibility." *The Journal of Philosophy* 66: 829–39.

Ginet, C. 1990. *On Action.* Cambridge: Cambridge University Press.

Hudson, H. 2006. *The Metaphysics of Hyperspace.* Oxford: Oxford University Press.

Krips, H. 1990. *The Metaphysics of Quantum Theory.* Oxford: Clarendon Press.

La Croix, R. 1976. "Omniprescience and Divine Determinism." *Religious Studies* 12: 365–81.

Lakatos, I. 1970. "Falsification and the Methodology of Scientific Research Programmes." In *Criticism and the Growth of Knowledge,* edited by I. Lakatos and A. Musgrave, 91–195. Cambridge: Cambridge University Press.

Leftow, B. 2012. *God and Necessity.* Oxford: Oxford University Press.

Lewis, D. 1983. "Survival and Identity." In *The Identities of Persons,* edited by A. O. Rorty, 17–40. Berkeley: University of California Press.

Lewis, D. 1986. *On the Plurality of Worlds.* Oxford: Blackwell.

McCall, S. 1994. *A Model of the Universe: Space-Time, Probability, and Decision.* Oxford: Clarendon Press.

Perszyk, K., ed. 2011. *Molinism: The Current Debate.* Oxford: Oxford University Press.

Rowe, W. 2004. *Can God Be Free?* Oxford: Oxford University Press.

Smart, J. J. C. 1963. "Free Will, Praise and Blame." *Mind* 70: 291–306.

Swinburne, R. 1977. *The Coherence of Theism.* Oxford: Oxford University Press.

Tooley, M. 1997. *Time, Tense and Causation.* Oxford: Clarendon Press.

Vaidman, L. 2014. "Many-Worlds Interpretation of Quantum Mechanics." In *The Stanford Encyclopedia of Philosophy*, edited by E. N. Zalta. http://plato.stanford. edu/entries/qm-manyworlds/.

Van Inwagen, Peter. 1975. "Incompatibility of Free Will and Determinism." *Philosophical Studies* 27: 185–209.

Woo, Chi. "Infinite Product Measure (Version 10)." http://planetmath.org/sites/ default/files/texpdf/38532.pdf.

4 An Argument for Modal Realism

Jason L. Megill

In this essay, I formulate an argument for a weak version of modal realism. To be precise, I argue that there are multiple (i.e., at least two) worlds that contain concrete entities. I conclude by discussing some implications the argument has for theism.

PRELIMINARIES

Before formulating the argument, I first define some key terms and explain precisely what it is the argument will attempt to show.

The distinction between 'concrete' entities, such as a particular chair, and 'abstract' entities, such as the concept 'chair', is commonplace in philosophy. For example, Rosen (2012, Section 1) states:

> It is widely agreed that the distinction [between concrete and abstract entities] is of fundamental importance . . . Thus it is universally acknowledged that numbers and the other objects of pure mathematics are abstract (if they exist), whereas rocks and trees and human beings are concrete. Some clear cases of abstracta are classes, propositions, concepts, the letter 'A', and Dante's *Inferno*. Some clear cases of concreta are stars, protons, electromagnetic fields, the chalk tokens of the letter 'A' written on a certain blackboard, and James Joyce's copy of Dante's *Inferno*.

And Lewis (1986, 82) remarks:

> A spectator might well assume that the distinction between 'concrete' and 'abstract' entities is common ground among contemporary philosophers, too well understood and uncontroversial to need any explaining.

Lewis expresses some misgivings about the usefulness of the distinction (see Lewis 1986, Section 1.7), but eventually decides to use it all the same.

This distinction between abstract and concrete entities plays an important role in debates about the ontological status of possible worlds. Many claim that our possible world, the actual world, is the only world that contains concrete entities; even though there are other possible worlds aside from the actual world, nothing in these worlds is concrete; the entities in these worlds are merely abstract. This view is sometimes called 'ersatzism'.[1] Ersatzism generally claims that possible worlds other than our own are representations of some sort. Indeed, there are various versions of 'ersatzism', which correspond to different possible mediums of representation: Carnap (1947) held that possible worlds are maximally consistent sets of *sentences,* while Plantinga (2003) holds that possible worlds are maximally consistent sets of *propositions.* Other versions include 'pictoral ersatzism', 'combinatorialism', 'nonreductivism', and 'fictionalism'.[2] However, other philosophers, most famously Lewis (1986), have endorsed 'modal realism', the claim that there are possible worlds aside from the actual world that contain concrete entities.[3] So, the debate about the ontological status of possible worlds has largely been a debate between two rival views, (1) ersatzism, which claims that the actual world is the only world that contains concrete entities and other possible worlds are merely abstract representations of ways that worlds could be, and (2) modal realism, which claims that there are possible worlds in addition to the actual world that contain concrete entities.

However, sometimes ersatzism might speak of entities being 'concrete' in some other possible world, which could create confusion: if we define 'ersatizsm' as the view that entities in other possible worlds are abstract and not concrete, then how can ersatzism claim that merely possible entities are 'concrete'? This confusion can be dissolved by drawing a distinction between two different ways an entity might be 'concrete'. Consider a possible world w that is not the actual world. w is the sort of possible world a defender of ersatzism would posit: it is just a representation of some sort, whatever the medium of representation might be (propositions, sentences, etc.). The ersatzer can say things like "entity e is concrete in w", but this will just mean something like, "e is represented as being concrete in w", or "*if* w were actual, e would be concrete". But e will not be thought of as being concrete *in the same way* that the concrete entities in the actual world are. So to avoid confusion, call entities that are concrete in the same way the entities in the actual world are 'literally concrete,' and call entities that are merely represented as being concrete 'representationally concrete'. In short, modal realism holds that there are entities in other possible worlds that are literally concrete, while ersatzism holds that there are entities in other possible worlds that are representationally concrete. Sometimes, for convenience, I will use the word 'CONCRETE' (in all capital letters) to denote literal concreteness; if an entity is CONCRETE, then it exists in exactly the same way that the concrete entities in the actual world do.

In the following, I formulate an argument for modal realism. But note that the conclusion of the argument is, relatively speaking, somewhat modest: there are at least two worlds that contain literally concrete enti-ties. Of course, this falls far short of, say, Lewis's robust version of modal realism, in which a number of claims are made about what worlds exist, and so on; for example, Lewis (1986) endorses 'plenitude', the principle of recombination, counterpart theory, the claim that these worlds are spatio-temporally isolated and the claim that they do not causally interact, and so on. The argument given in the following cannot (and does not try to) establish the truth of these other claims. If one prefers, one could distinguish between three positions: (a) ersatzism (there is only one possible world—ours—that contains literally concrete entities), (b) Lewis's modal realism (every logically possible world will be literally concrete), and (c) 'restricted modal realism' (which only claims that there are multiple—at least two—possible worlds that contain literally concrete entities). The argument given later—if successful—would show that (a) is false and that (c) is true. Given (c), (b) might also be true, but the argument does not show that (b) is true.[4]

Finally, I will sometimes speak of 'literally concrete possible worlds', which are worlds (such as our own) that contain literally concrete entities. But note that calling a world a 'literally concrete possible world' does not imply that *all* entities in that world are concrete. Abstract entities might exist in the world as well. For example, the actual world is a literally con-crete possible world in that it contains some literally concrete entities; but abstract entities (numbers, concepts, whatever one thinks is abstract, etc.) might exist (in whatever sense in which they exist) in the actual world as well. The fact that a world contains some literally concrete entities does not preclude it from containing abstract entities too. And the presence of liter-ally concrete entities does not entail that abstract objects exist either; so of course the notion of a 'literally concrete possible world' is consistent with skepticism about abstract objects (e.g., nominalism) as well.

In sum, I will argue that there are at least two possible worlds that con-tain literally concrete entities. This claim is inconsistent with ersatzism but is more modest than Lewis's modal realism.

THE ARGUMENT

I now formulate an argument for modal realism. Premise one is:

(1) *If* an entity e is possibly literally concrete in the actual world, *then* there is a possible world w in which it is literally concrete.[5]

Note that this claim is consistent with both modal realism and ersatzism, and so does not beg the question against ersatzism. One might object that while the world in the antecedent of (1) is the actual world, perhaps the world w in the consequent is a different world; if so, then there are literally concrete objects in at least two worlds, so (1) is inconsistent with ersatzism. But note that it might be that the actual world and world *w* are always the same world; that would be consistent with (1). And in this case, there would be only one world that contains literally concrete entities, which is consistent with ersatzism. In short, (1) is logically consistent with both ersatzism and modal realism.[6]

Premise Two is:

(2) There (i) is an entity *e* that is possibly literally concrete in the actual world, but (ii) *e* is not literally concrete in the actual world.

This premise claims that not all entities that are possibly literally concrete in the actual world are literally concrete in the actual world; there are entities that could be literally concrete in the actual world, but are not. Note that (2) does not claim that *e* is literally concrete in a world; it only claims that *e* is possibly literally concrete in our world.[7] (2)—in isolation—is consistent with *e* simply being an abstraction of some sort, or simply being representationally concrete; that is, like (1), (2) is consistent with ersatzism and so does not beg the question.[8]

Both premises are defended later, but if they are true, then,

(3) There are multiple (i.e., at least two) possible worlds that contain literally concrete entities.[9]

For suppose that (1) and (2) are true. Premise (1) claims that *if* an entity *e* is possibly literally concrete in the actual world, *then* there is a possible world *w* in which it is literally concrete. Premise (2) claims that there (i) is an entity *e* that is possibly literally concrete in the actual world yet (ii) *e* is not literally concrete in the actual world. Given (2), and so (i), there is an entity *e* that is possibly literally concrete in the actual world. But then, given (1), there is a possible world *w* in which *e* is literally concrete. But also note that, given (2), and so (ii), this possible world *w* in which *e* is literally concrete cannot be the actual world (i.e., *the actual world ≠ w*). Therefore, *e* must reside in some literally concrete world that is not the actual world. That is, there are multiple (or at least two) literally concrete possible worlds; our world and *w*. So, (1) and (2) entail (3); the argument is valid.[10] But is the argument sound? If (1) and (2) are true, it is. I defend premise (2) in the following section and premise (1) in the section thereafter. In my estimation, the second premise is undoubtedly true; so after showing this, I can focus on what I take to be the only real issue concerning the argument's soundness: the truth of the first premise.

A DEFENSE OF PREMISE TWO

Again, premise (2) is: "there (i) is an entity *e* that is possibly literally concrete in the actual world but (ii) *e* is not literally concrete in the actual world".

Quite simply, there are entities that could have been CONCRETE in the actual world but are not. In other words, there are possible worlds that contain entities that our world does not, yet these worlds are otherwise just like the actual world. So, for example, there is a grain of sand that is not CONCRETE in our world, but it possibly could have been; so there is a possible world that is just like the actual world except that it contains this grain of sand (whether this grain of sand is merely represented as existing in this other world or not). In short, not everything that could have been CONCRETE in our world is.

I first offer a few reasons for thinking that (2) is *prima facie* plausible, before offering three more detailed arguments for it. *First,* note that (2) posits an entity that is possibly CONCRETE in our world, but is not CONCRETE in our world; (2) does not claim, however, that this entity *is* CONCRETE in some other world (claiming that would beg the question against ersatzism). (2) is consistent with this entity being merely representationally concrete. Again, (2) is consistent with ersatzism. And indeed, both modal realists and ersatzers think that there are entities that could have been in our world (or in a world that is otherwise just like ours) but are not, so few on either side of the debate will be inclined to reject (2). *Second,* consider the so-called 'Copernican Principle', the idea that humans do not occupy a privileged place in the cosmos; there is nothing particularly special about our location in the universe. One might posit an 'Extended Copernican Principle', and hold that there is nothing particularly special about our universe as a whole and the particular objects in it; it is not as if the concrete entities in our universe are special in that they and only they could have been concrete in our world. *Third,* it could be that those who deny (2) shoulder the burden of proof. (2) posits a logically possible entity (specifically, an entity that is possibly literally concrete in the actual world but is not); and, one could argue that those who deny that something is logically possible (and so, for example, deny that the entity posited by (2) is possible) must shoulder the burden of proof.[11] That is, arguably, the default position should be that (2) is true until it is proven otherwise. Nevertheless, I now give three more systematic arguments for (2).

Argument One

Note that,

(A) If (2) is false, then everything that is possibly literally concrete in the actual world is literally concrete in the actual world.

Claim (A) is clearly true. Here is a proof of (A): suppose, for conditional introduction, that (2) is false. (2) is a conjunction: "there (i) is an entity *e* that is possibly literally concrete in the actual world but (ii) *e* is not literally concrete in the actual world". But we know that (i) is true: there are many literally concrete entities in the actual world, so there are many entities that are possibly literally concrete in the actual world. We know that there are many entities that are possibly CONCRETE in the actual world because there are many CONCRETE entities in the actual world.[12] So, if (2) is false, (ii) must be false; again, (2) is a conjunction, and we know that one conjunct (i.e., (i)) is true, so if (2) is false, the other conjunct (i.e., (ii)) must be false. That is, there are not any entities that could have been literally concrete in the actual world that are not literally concrete in the actual world. Everything that is possibly CONCRETE in the actual world is CONCRETE in the actual world. (A) is a conditional; we were able to assume the antecedent of (A) and derive the consequent; so (A) is true.

However, it appears that,

(B) Not everything that is possibly literally concrete in the actual world is literally concrete in the actual world.

If (B) is false, then there are no entities that could be literally concrete in the actual world but are not. Every entity that is possibly literally concrete in the actual world is; and if an entity is not literally concrete in the actual world, it could not possibly be literally concrete in the actual world. But why would this be the case? What mysterious property do the literally concrete entities in the actual world have that makes them—and only them—possibly literally concrete in the actual world? Why would a grain of sand that is only slightly different than one that already literally concretely exists (in the actual world) not be possibly literally concrete (in the actual world)? In short, a denial of (2) seems wholly *ad hoc* and mysterious. And intuitively, one could give many examples of entities that could have been CONCRETE in the actual world, but for whatever reason, or perhaps even for no reason at all, are not CONCRETE in the actual world. For example, in 1956, Frank Lloyd Wright designed a skyscraper to be built in Chicago called 'The Illinois'. The Illinois was never constructed and so never made CONCRETE, but surely it could have been constructed. Indeed, a denial of (2) commits one to various claims that have few defenders in contemporary metaphysics. For instance, if (2) is false, then there are no possibly literally concrete entities in the actual world that are not literally concrete in the actual world; so if something is possibly CONCRETE in the actual world, it is CONCRETE in the actual world. This echoes claims made by Spinoza ([1677] 1985), claims that are widely rejected today.[13] In sum, our intuitions about what is possible suggest that (B) is true; these intuitions are powerful and widely shared.

But if (A) and (B) are true, then:

(C) (2) is true.

(C) follows from (A) and (B) by modus tollens.

Argument Two

One common way of characterizing what is *logically* possible is to claim that "if *p* is not logically contradictory, then *p* is logically possible". See, for example, Szabo-Gendler and Hawthorne (2002, 4): "on a standard sort of characterization, *P* is logically possible just in case no contradiction can be proved from *P* using the standard rules of deductive inference . . ." If the logically consistent is also the logically possible, then:

(A) *If* (there (i) is an entity *e* such that, if it were added to the actual world, no contradiction would result, even though (ii) *e* is not CONCRETE in the actual world), *then* (there (i) is an entity *e* that is possibly CONCRETE in the actual world but (ii) *e* is not CONCRETE in the actual world).[14]

Moreover:

(B) There (i) is an entity *e* such that, if it were added to the actual world, no contradiction would result, even though (ii) *e* is not CONCRETE in the actual world.

Consider, for example, a particular grain of sand *s* that is CONCRETE in the actual world. Now, consider a grain of sand *s'* that has the same intrinsic properties as *s*; *s'* is qualitatively identical to *s*. Would the addition of *s'* into the actual world generate a contradiction? There appear to be two ways that the introduction of an entity into a world could generate a contradiction: (i) the entity is self-contradictory (like, e.g., a square circle is) or (ii) the entity is not self-contradictory, but a contradiction is generated when it is combined with some entity (or collection of entities) external to itself.[15] But we know that *s'* is not self-contradictory: it is a duplicate of an already existing grain of sand. If *s'* is self-contradictory, then *s* is too since they are duplicates; but by hypothesis, *s* literally concretely exists (in the actual world) and so cannot be self-contradictory; therefore, *s'* cannot be self-contradictory either. And we know that *s'* cannot generate a contradiction when combined with any other entity (or collection of entities) in the actual world: *s* does not generate a contradiction when combined with any other entity (or collection of entities) in the actual world (again, *s* exists in the actual world); and *s* and *s'* are duplicates, therefore, *s'* must not generate a contradiction when combined with any other entity (or collection of entities) in the actual world either. There are only two ways that *s'* could generate a contradiction in the actual world; but *s'* would not generate a contradiction in either of these ways. So, *s'* could consistently exist as a literally concrete entity in a world that is otherwise just like the actual world. (B) is true.

But given (A) and (B), we can infer with modus ponens that:

(C) There (i) is an entity *e* that is possibly CONCRETE in the actual world but (ii) *e* is not CONCRETE in the actual world.

And (C) is simply premise (2) of the argument for modal realism. (2) is true.[16]

Argument Three

Here is a third argument for premise (2).[17] This argument depends on the notion of a 'unique' entity. If an entity is 'unique', then there is no other entity in the world that has the same exact set of intrinsic properties; a unique entity is not qualitatively identical to any other entity in the world; a unique entity currently lacks a CONCRETE twin. The first step is:

(A) If a unique entity *d* is CONCRETE in the actual world, then a numerically distinct duplicate of *d*—an entity *e* that is not *d* but is exactly like *d* in all (intrinsic) respects—is possibly CONCRETE in the actual world.

For if we know that a unique entity *d* already exists in the actual world, and so is in fact CONCRETE in the actual world, then we know that entities that are just like *d* are possibly CONCRETE in the actual world. So, another entity *e*—that is numerically distinct from yet qualitatively identical to *d*— should be at least be possibly CONCRETE in the actual world. For example, consider a particular rock in my yard. Given that this rock is CONCRETE in the actual world, it should be possible for another rock that is (intrinsically) just like it to be CONCRETE in the actual world too. It should be possible for this rock to have a twin. For anything that would prevent this twin rock from being possibly CONCRETE in our world would prevent the original rock from being possibly CONCRETE in our world as well; yet the original rock is CONCRETE in our world, and so is possibly CONCRETE in our world. Therefore, the twin rock is possibly CONCRETE in our world as well.[18]

And clearly,

(B) There is a unique entity *d* that is CONCRETE in the actual world.

(B) simply states that there is at least one CONCRETE entity *e* in the actual world that is 'unique'. Examples are all around us; for example, this rock in my yard. It is extremely unlikely that there is another rock in the actual world that is exactly the same as this rock in my yard; the twin rock would need to be exactly the same shape, comprise the same number of atoms, and so on. Also note that (B) is true if there is just a *single* entity in the actual world that lacks a concrete twin; this appears highly probable. In any event, there are some entities in the actual world that we *know* lack a concrete twin in the actual world, for example, the Sears Tower. But if (A) and (B) are true, then we can infer with modus ponens that:

(C) A numerically distinct duplicate of *d*—an entity *e* that is not *d* but is exactly like *d* in all (intrinsic) respects—is possibly CONCRETE in the actual world.

So, given (C), *e* is possibly CONCRETE in the actual world. But, given (B), *e* is not CONCRETE in the actual world (again, (B) claims that *d* is 'unique' in the actual world in that it lacks a CONCRETE duplicate in the actual world, so *e* cannot be in the actual world). So:

(D) There (i) is an entity *e* that is possibly literally concrete in the actual world but (ii) *e* is not literally concrete in the actual world.

And (D) is premise (2) of the argument for (restricted) modal realism.

In sum, there is ample reason to endorse (2): there are entities that could have concretely existed alongside the CONCRETE entities in our world, but do not. But if the argument for modal realism is valid and (2) is true, then if (1) is also true, the argument is sound. I now defend (1).

A DEFENSE OF PREMISE ONE

Again, premise (1) is "*if* an entity *e* is possibly literally concrete in the actual world, *then* there is a possible world *w* in which *e* is literally concrete". That is, if *e* could have been CONCRETE in the actual world, then there is a world in which *e* is CONCRETE. In this section, I defend (1) with three distinct arguments.

Argument One

In possible worlds semantics, if something is logically possible, then this simply means that there is a logically possible world in which it is the case. See, for example, Parent (2012, Introduction): "'Possibly, *p*' is said [in the most elementary kind of Kripkean logic] to be true if, and only if, there is at least one possible world in which the state-of-affairs *p* obtains". Kripke's formulation of possible world semantics (see, e.g., Kripke 1963) was of course a vital development in modal logic, "a great advance in logic", as Parent (2012: Introduction) claims. So, (A) should not be controversial:

(A) If *p* is logically possible, then there is a logically possible world *w* in which *p* obtains.

Moreover, consider what can be called a 'collection of facts'. A collection of facts is simply a group or set of facts, whatever the facts might be. Some collections of facts will be logically possible; it will be logically possible that these facts all obtain at once. Other collections of facts will not be logically possible, for example, contradictory collections of facts. But given (A), we can infer that:

(B) If a collection of facts is logically possible, then there is a logically possible world *w* in which this collection of facts obtains.

(B) is simply a substitution instance of (A), so it is not controversial either.[19] Now, assume for conditional introduction that:

(C) There is an entity *e* that is possibly literally concrete in the actual world.

(C) is an assumption for conditional introduction and so needs no justification. But as (C) asserts, one fact about *e* is that it is possibly CONCRETE in our world. So, the fact that *e* could be literally concrete will combine with other facts about *e*, whatever they might be, and thereby form a collection of facts that is logically possible. To explain, there will be various facts about *e*; it is red, it has four wheels and so on. These facts will combine to form a logically possible collection of facts. But if *e* is also possibly CONCRETE, then the fact that *e* is CONCRETE can combine with this previous collection of facts to form a new—also logically possible—collection of facts. So, for example, suppose that there is a logically possible car that is red, has four wheels, is fast, and so on. These facts about this car will form a logically possible collection of facts. But also suppose that the car could be yellow instead of red. Now we have a second logically possible collection of facts; all of the same facts will be true about this car that were true before, except now the fact that it is yellow will hold instead of the fact that it is red. Likewise, we could take our initial collection of facts and add the following fact: the car is literally concrete. So now we have a third logically possible collection of facts: the car is red, has four wheels, is fast, and is CONCRETE. So, given (C), that is, given that there is an entity *e* that is possibly literally concrete in the actual world, we can infer:

(D) There is a collection of facts—in which *e* is literally concrete—that is logically possible.

But then we can infer, with modus ponens and (B) and (D), that:

(E) There is a logically possible world *w* in which this collection of facts—one of which is that *e* is literally concrete—obtains.

And if so, then:

(F) There is a possible world *w* in which *e* is literally concrete.

If (E) is true, then there is a possible world in which various facts obtain, and one of these facts is that *e* is CONCRETE. Indeed, (F) follows from (E). Finally, with conditional introduction on (C) through (F), we can infer that:

(G) If an entity *e* is possibly literally concrete in the actual world, then there is a possible world *w* in which *e* is literally concrete.

Claim (A) is a core tenet of possible worlds semantics. Claim (B) is merely a substitution instance of (A). (C) is an assumption for conditional introduction and so can't be denied; (D) follows from (C); (E) follows from (B) and (D) with conditional elimination; (F) follows from (E); and (G) follows from (C) though (F) with conditional introduction and so can't be denied, at least given the truth of the other steps. Of course, (G) is premise (1) of the argument for modal realism. So (1) is true.

Perhaps an example can clarify the argument. Consider again a particular car. There will be various facts about this car: it is red, it has four wheels and so on. Suppose also that this car is possibly literally concrete in the actual world. This car could have been CONCRETE in our world. So, there will be a logically possible collection of facts in which this car is literally concrete. This collection of facts, since it is logically possible, will obtain in a logically possible world. But then the car is concrete in this world, and so is concrete in a world. If the car is possibly concrete in the actual world, it will be literally concrete in some world or other. But then (1) is true.

To further clarify: suppose (as some versions of ersatzism claim) that a world *w* is, for example, a set of propositions *S*; *S represents w* as being a certain way. Certain facts are represented as being true in *w*, others are not. These propositions in *S* describe a consistent collection of facts, and so a logically possible world. But these propositions *could* all obtain in a world that is literally concrete. For one *could* take *S*, and then combine it with the fact that the propositions in *S* obtain in a world that is actualized. This would then produce a *new* consistent collection of facts, *S'*, one in which *S* describes an actualized, literally concrete world. And so this world will be logically possible as well. And if premise (2) is also true, this concrete world will not be the actual world, and so there will be at least two worlds that contain CONCRETE entities.

To clarify still more: consider, for example, step (F): "there is a possible world *w* in which *e* is literally concrete". The defender of ersatzism will likely respond, "if *w* is not the actual world, then *e* is not *literally* concrete in *w*. It's simply that *e* is *represented* as being concrete in *w*, whatever the medium of representation might be (i.e., whichever version of ersatzism is true). For example, perhaps worlds are simply sets of propositions, and in *w*, the proposition '*e* is concrete' is true, that is, the proposition *represents* (and only represents) that *e* is concrete". But this objection misses the point of the argument. The argument is claiming that there might be, say, a consistent set of propositions that represent a world as being a certain way. This set of propositions will describe a consistent set of facts, a possible collection of facts, a possible world. But then—in turn—*this* possible collection of facts *could be* combined with an *additional* fact, namely, that the initial collection of facts holds in a world that is *literally* concrete; this would thereby produce a world in which the propositions refer to entities that are CONCRETE. Essentially, one can posit a possible world *w* that is acceptable to ersatzism, that is, a world in which entities are merely representationally concrete. But if it is also possible that this world is literally concrete, then the fact that it is literally concrete can be combined with all of the other facts in *w*, thereby producing a new world *w** that is literally concrete. So even if one begins by assuming ersatzism, it can be shown that ersatzism is false.

Finally, one might object: "either (i) *e* is merely possibly CONCRETE or else (ii) you are stipulating that *e* refers to a CONCRETE object that is not actual. But if (i), *e* can only be a *merely possibly CONCRETE object;* such an object poses no threat to ersatzism. But if (ii), then you are assuming that *e* is CONCRETE and so are begging the question against ersatzism". This objection misconstrues the structure of the argument. Premise (2)—taken in isolation—is consistent with *e* being merely possibly CONCRETE. But when premise (2) is combined with premise (1), we see that *e* must be CON-CRETE and not merely possibly CONCRETE. But it is not *stipulated* that *e* is CONCRETE. Rather, the argument for premise (1) attempts to show that if *e* is possibly CONCRETE, then it is CONCRETE in some world or other. So, the argument does not beg the question.

Argument Two

Consider all logically possible worlds. Any way that a world can be corresponds to one of these logically possible worlds. The collection of all logically possible worlds will be *complete;* if a world can be a certain way, then in the collection of all logically possible worlds, there will be a world that is that way. For it would be incoherent to claim both that (a) it is logically possible for a world to be a certain way yet (b) there are no logically possible worlds that are way.[20] In sum:

(A) If it is logically possible for a world to be such that *p* obtains, then there is a logically possible world in which *p*.[21]

Now suppose, for conditional introduction, that:

(B) An entity *e* is possibly literally concrete in the actual world.

(B) claims that there is an entity that could have been literally concrete in the actual world. In other words, there is a logically possible world that is just like the actual world except that it contains a literally concrete *e*. That is, another way of saying (B) is:

(C) It is logically possible for a world to be such that *e* is literally concrete.

But note that the following, that is, (D), is merely a substitution instance of (A) (in which we substitute "*e* is literally concrete" for *p*):

(D) If it is logically possible for a world to be such that *e* is literally concrete, then there is a possible world *w* in which it is literally concrete.

Moreover, given (C) and (D), we can infer with modus ponens that:

(E) There is a possible world *w* in which it (i.e., *e*) is literally concrete.

Finally, given (B) through (E), we can infer with conditional introduction that:

(F) If an entity e is possibly literally concrete in the actual world, then there is a possible world w in which it is literally concrete.

Of course, (F) is premise (1) of the argument for (restricted) modal realism. Step (A) seems fairly trivial; (B) is an assumption for conditional introduction and so needs no justification; (C) is simply another way of saying (B); (D) is a substitution instance of (A) and so is true if (A) is; and (E) and (F) follow with basic rules of logic (conditional elimination and conditional introduction, respectively).

The point of the argument is this: it appears that the space of logical worlds is complete; every way that a world can be corresponds to the way that a world is. But one way that worlds (even those other than our own) could be is this: they could be CONCRETE. So ersatzism effectively posits gaps in the space of logically possible worlds; some worlds could have been CONCRETE, and so these CONCRETE worlds should be in the space of logically possible worlds, but ersatzism denies that there is a corresponding world in the space of logically possible worlds in which they are CONCRETE. Ersatzism claims that some worlds that are in fact in the space of logically possible worlds are not.

One might object: ersatzism does claim that these worlds are in the space of logically possible worlds. They are there; they are just merely representationally concrete. To ersatzism, these worlds are merely abstract possibilities, and so are not CONCRETE. Saying that a world is CONCRETE merely means that it is represented as being so; it need not be literally concrete. But this objection misses the point: if these worlds could in fact be CONCRETE, they should be CONCRETE in the space of logically possible worlds, *contra* ersatzism. If an abstract representation could be actual, or is possibly actual, and if all possibilities obtain somewhere in the space of logically possible worlds, then somewhere in the space of logically possible worlds, the states of affairs that these abstract representations represent should be actual.

Argument Three

Suppose that possible worlds (aside from the actual world) are merely abstractions of some sort. Assume, for instance, that possible worlds are simply maximally consistent sets of propositions (see, e.g., Plantinga 2003), as again, one influential version of ersatzism claims. And consider the following example: suppose that Joe has a yellow car in the actual world. Joe considered painting his car red, but decided not to. But also suppose that Joe might have decided differently; he might have decided to paint his car red. So his car is possibly red, even though in the actual world it remains yellow. So, for example, consider a possible world just like ours, except in this world, Joe decided to paint his car red. Call this world 'w'. Again, we have assumed a particular form of ersatzism: w will be a consistent set of propositions, such as "Joe's car is red", "Bush won the 2000 election", and so on. In w, all of these propositions will be true.[22] But, relative to the actual world, can w be possibly literally concrete yet not literally concrete? If this world is possibly literally concrete relative to the actual world, then the proposition "this world is literally concrete" will be true in w. But if "this world is literally concrete" is true in w, w must be literally concrete. It is not possible for a world (or an entity) to be possibly literally concrete yet not literally concrete. (1) is true.

Here is a slightly different and more systematic version of the argument. Assume, for conditional introduction, that:

(A) e is possibly literally concrete in the actual world.

As an assumption for conditional introduction, (A) needs no justification. Moreover, if e is possibly literally concrete in the actual world, then there is a possible world, which we can call 'w', in which e is literally concrete; this follows from the definition of 'possibility' in possible worlds semantics. And for the purpose of this argument, it is important to interpret this claim in such a way that it is clearly consistent with ersatzism; *we must be careful not to beg the question against ersatzism.* So, suppose that worlds aside from our own are abstract entities; for example, suppose that worlds are simply maximally consistent sets of propositions, as one influential version of ersatzism claims. So, when we say that "there is a possible world w in which e is literally concrete", this simply means that the proposition "e is literally concrete" is true in w. That is,

(B) If e is possibly literally concrete in the actual world, then there is a possible world w in which (the proposition) "e is literally concrete" is true.

The point is that even ersatzism—or at least the prominent version of it we are assuming here—should have no issue with (B). Indeed, it's not merely that this version of ersatzism will have no problem with (B); rather, this version of ersatzism *entails* (B). If a possible world is simply a set of propositions, and so saying that something is possible (e.g., e is possibly literally concrete) is simply saying that there is a world where the relevant proposition is true (e.g., "e is literally concrete" is true), then (B) *must* be true. But given (A) and (B), (C) follows with modus ponens:

(C) There is a possible world w in which (the proposition) "e is literally concrete" is true.

But note that:

(D) If "e is literally concrete" is true (in w), then e is literally concrete (in w).

For if *e* is not literally concrete (in *w*), then "*e* is literally concrete" will be false (in *w*). To deny (D), one has to claim that a proposition is true but the state of affairs which it posits does not obtain; this is incoherent. But given (C) and (D), we can infer with modus ponens that:

(E)   *e* is literally concrete (in *w*).

But then, with conditional introduction and (A)—(E), we can infer that:

(F)   If *e* is possibly literally concrete in the actual world, then *e* is literally concrete in *w*.

Of course, (F) is premise (1) of the argument for modal realism. This argument consists of an assumption (A) for conditional introduction that needs no justification; a claim (B) that even ersatzism accepts and indeed entails; a claim (C) that follows from (A) and (B) with modus ponens; a claim (D) whose denial is incoherent; a claim (E) that follows from (C) and (D) with modus ponens; and a conclusion (F) that follows from (A)—(E) with a basic logical inference.

One might object that the argument, at least as it is formulated earlier, only applies to one form of ersatzism; perhaps other forms of ersatzism can overcome the argument. But note that the argument could be adapted to apply to any version of ersatzism. All forms of ersatzism claim that possible worlds are merely abstract representations of a world, whether the medium of representation is sentential, propositional, pictoral or something else; see, for example, Parent (2012, Section 3): "This indicates another shared feature of worlds among ersatzers; a world-surrogate is in some sense representational". All of these views will represent that "*e* is literally concrete" in some fashion or other, so the argument can be adapted to apply to any of them (simply replace 'propositions' with whatever medium of representation a given form of ersatzism appeals to).

The opponent of modal realism will likely deny (D); indeed, (D) is arguably the only step that the proponent of ersatzism *can* deny. Again, D claims, "If '*e* is literally concrete'" is true (in *w*), then *e* is literally concrete (in *w*)". The proponent of ersatzism will likely claim that the proposition "*e* is literally concrete in *w*" represents *e* as being literally concrete in *w*, and so the proposition is true in *w*, but *e* is not literally concrete in *w*. But again, this appears incoherent, for the proposition will then be both true (because its truth represents that the state of affairs it describes obtains in *w*) and false (because *e* is not in fact literally concrete in *w*).

CONCLUDING REMARKS

I conclude by considering some implications the argument has for theism. There are certain logical relationships between the argument for (restricted) modal realism and multiverse theories in physics. First, note that if the argument for modal realism given earlier is sound, then there is a multiverse. Assume that there are at least two worlds that contain literally concrete entities. And note that there are two possibilities: either (i) a possible world can contain only one universe (because, e.g., a possible world simply is a universe . . . 'possible world' and 'universe' are equivalent terms) or (ii) not, in which case a possible world can contain more than one universe, that is, a multiverse. If (i) is true, that is, if a possible world just is a universe and vice versa, then given that there are at least two concrete possible worlds, there are at least two universes, and so there is a multiverse. If (ii) is true, that is, if a possible world can contain a multiverse, then given that there are at least two CONCRETE possible worlds, there will be at least two concrete universes, and so there is a multiverse. Either way, if the argument for (restricted) modal realism is sound, there is a multiverse.[23]

To be more precise, again, one can draw a distinction between modal realism, which claims that every logically possible world is CONCRETE, and restricted modal realism, which only claims that there are at least two CONCRETE logically possible worlds. Modal realism entails restricted modal realism but not vice versa; if every logically possible world is CONCRETE, then at least two worlds are CONCRETE. But given that at least two possible worlds are CONCRETE, it might be that all logically possible worlds are CONCRETE and it might not be. Likewise, there are a number of extant multiverse theories in physics; see, for example, Tegmark (2003) for a taxonomy of these different theories. Some of these theories seem to cohere better with *just* restricted modal realism while others seem to cohere better with Lewis's robust modal realism. For example, some of these multiverse theories—like the many-worlds interpretation of quantum mechanics—claim that all CONCRETE universes will have the same physical constants as our universe. But if so, then it appears that since there are logically possible universes with different physical constants, not all logically possible universes will be CONCRETE. This would be consistent with *mere* restricted modal realism; that is, a restricted modal realism that denies Lewis's robust modal realism, and so denies that all logically possible worlds will be CONCRETE. However, other multiverse theories are different. For example, Tegmark's (2003) theory claims that any universe that can be described in a mathematical structure is real (i.e., is CONCRETE). Tegmark's theory would seem to be analogous to Lewis's view, which claims that all logically possible worlds are CONCRETE.[24]

Of course, multiverse theories have important implications for theism and the philosophy of religion in general. For example, some have suggested that theism should posit a multiverse to overcome the problem of evil; for further discussion of the multiverse response to the problem of evil (see, e.g., McHarry 1978; Forrest 1981; Turner 2003; Almeida 2008; Kraay 2010; Megill 2011). Roughly, the idea is that while we are not in the best of all possible worlds, as Leibniz famously claimed, perhaps we are in the best of all possible multiverses. Advocates of the argument from evil ask why God didn't create a better universe than ours, one that contains less suffering. But if there is a multiverse, then perhaps God did create a better universe than ours, in addition to creating ours. There is a multiverse that contains our universe, along with better universes. Indeed, the multiverse will contain all possible universes that are worthy of creation. But note that the *composition* of the multiverse appears relevant: at least *prima facie,* it seems there are possible universes that God should not and would not actualize. There seem to be possible universes rife with gratuitous evil and so on. So, if a view like Lewis's or Tegmark's is true, then all logically possible universes will be CONCRETE; but then universes that are not worthy of actualization will be CONCRETE. This would undermine the multiverse response to the problem of evil, if not theism itself. Then again, if *only* a restricted modal realism is true, or if one of the multiverse theories in physics that do not claim that all possible universes will be CONCRETE is true, then the possibility remains that *only* worthy universes (or worlds) are CONCRETE, and so the multiverse response to evil *might still* work.

In short, the argument for restricted modal realism given earlier, if sound, can bolster the multiverse response to the problem of evil insofar as it suggests that there is in fact a multiverse. But any argument for a multiverse cannot be *too* strong, for if it entails that all logically possible universes are CONCRETE, the multiverse response to the problem of evil—and theism itself—is undermined. So, for the theist at least, the argument for restricted modal realism given earlier has two advantages: (i) it is strong enough to show that there is a multiverse, and so strengthens the multiverse response to evil, but (ii) it is not *too* strong, that is, it doesn't entail that all universes are CONCRETE, thereby undermining the multiverse response to evil.

A similar dynamic arises in connection with fine-tuning arguments for God's existence. It appears that a number of laws of nature and physical constants have to be just right for life—or at least human life—to be possible. That is, "almost everything about the basic structure of the universe . . . is balanced on a razor's edge for life to occur" (Collins 1999, 48). Consider, for example, the Big Bang. Scientists tell us that:

> If the initial explosion of the big bang had differed in strength by as little as one part in 10^{60}, the universe would have either quickly collapsed back on itself, or expanded too rapidly for stars to form. In either case, life would be impossible. (Collins 1999, 49)

Gravity provides another example: "if gravity had been stronger or weaker by one part in 10^{40}, then life-sustaining stars like the sun could not exist. This would most likely make life impossible" (Collins 1999, 49). Given just how improbable it would be for all of these conditions to align through pure chance, perhaps we should conclude that an intelligent being is behind the actualization of our universe?

Arguably, the most serious objection to fine-tuning arguments is the possibility that multiple universes exist. Himma (2006, Section 2) writes that

> some physicists speculate that this physical universe is but one material universe in a "multiverse" in which all possible material universes are ultimately realized. If this highly speculative hypothesis is correct, then there is nothing particularly suspicious about the fact that there is a fine-tuned universe, since the existence of such a universe is inevitable (that is, has probability 1) if every material universe is eventually realized in the multiverse. Since some universe, so to speak, had to win, the fact that ours won does not demand any special explanation.[25]

In short, just as the theory of evolution allows science to explain the apparent design of biological organisms without appeal to an intelligent agent, the multiverse hypothesis would allow science to explain the apparent fine-tuning of the laws of nature without appeal to an intelligent agent.

But again, the *type* of multiverse that one posits is relevant. It appears that if *all* logically possible worlds or universes exist, then the multiverse response to the fine-tuning argument would work. However, if only *some* logically possible worlds or universes are concrete, as a restricted modal realism (without Lewis's modal realism) or the many-worlds interpretation of quantum mechanics claims, then fine-tuning arguments might still work *even if* there is a multiverse. For example, as mentioned earlier, the many-worlds interpretation of quantum mechanics claims that all concrete universes will have the same physical constants as our universe. But then it is improbable that the physical constants in the *multiverse* are what they are, and the fine-tuning argument can still go through. In short, for the theist, at least, the argument for modal realism given earlier has an advantage: it does not show that *all* logically possible worlds or universes are concrete, but rather only that *some* are, and so the argument leaves open the possibility that fine-tuning arguments for God's existence might work.[26]

NOTES

1. See, e.g., Parent (2012, Section 3): "Most basically, the Ersatzer construes talk about a possible world as talk about some *ersatz* object. ('Ersatz' is German for 'replacement' or 'substitute'.) Thus the truth or falsity of a modal statement is explained by appeal to surrogates or proxies for possible worlds, rather than to genuinely existing worlds themselves".
2. See Parent (2012) for a discussion of these various forms of ersatzism.
3. See Parent (2012, Section 2): Lewis holds that "there is only one kind of being, and all possibilia . . . have it. Thus Lewis provocatively suggests that non-actual possibles [or possibles that are non-actual in our world] exist in just the same way that you and I do" (1986, 2–3). Others who have defended modal realism include Richard Miller (1993, 2001), Jackson, Priest, and McDaniel (2004), and Daly (2008).

4. There are several obvious logical relationships between (a), (b), and (c): (a) is inconsistent with (b) and (c); (b) entails (c) but (c) taken by itself is consistent with both the truth and falsity of (b). I discuss this further in the concluding remarks.

5. Throughout, I am concerned with logical possibility as opposed to, say, nomological possibility.

6. In case this is unclear, it might be that in all cases, the world w in the consequent is numerically identical to the actual world; (1) is consistent with this. And this would be consistent with ersatzism, for then there would only be one world that contains literally concrete entities. Or, more controversially, it might be that there are *also* some cases in which the actual world and w are numerically distinct worlds, in which case ersatzism will be false (since there will be multiple worlds that contain literally concrete beings). (1) is consistent with both possibilities, and so is consistent with ersatzism and modal realism. I belabor the point to obviate any potential worries that (1) is inconsistent with ersatzism and so begs the question.

7. And in case it isn't clear, when I say that "*e* is possibly literally concrete in our world," I mean that there is a world that is just like our world except that it also contains *e*.

8. Both (1) and (2) are—taken individually—consistent with ersatzism. Of course, they are jointly inconsistent with ersatzism, but that is not begging the question. Indeed, I argue for the truth of both (1) and (2) below; assuming that these arguments succeed, then *even if* (1) or (2) *is* individually inconsistent with ersatzism, ersatzism would be false in any event. A claim that is inconsistent with a true claim is false.

9. One could object: "a world might contain literally concrete entities yet might not be concrete . . . a set might contain a literally concrete entity as a member (e.g., the set of rocks in my yard) yet sets are not concrete". But one might wonder if the analogy is strong: is set membership analogous to an entity being located in a world in such a way that this objection has merit? Moreover, note that the actual world is a world that contains literally concrete entities and also (presumably) abstract entities. The interesting question is whether there are other logically possible worlds aside from the actual world that also contain literally concrete entities (as well as perhaps abstract entities). That is, is the actual world unique in being the only world that contains literally concrete entities? If the argument for (restricted) modal realism given above is sound, this question would be answered.

10. If one harbors any doubts about the argument's validity, suppose that the premises are true:

(A) *If* an entity e is possibly literally concrete in the actual world, *then* there is a possible world w in which it is literally concrete.

(B) There (i) is an entity e that is possibly literally concrete in the actual world yet (ii) e is not literally concrete in the actual world.

Given (B), (i) follows with conjunction elimination, i.e.:

(C) There is an entity e that is possibly literally concrete in the actual world.

But then, with (A) and (C), conditional elimination entails:

(D) There is a possible world w in which e is literally concrete.

But given (B), we can infer (with conjunction elimination) that:

(E) e is not literally concrete in the actual world.

But since (D) e is literally concrete in a world w and (E) e is not literally concrete in the actual world, then:

(F) The actual world and w cannot be the same world. That is, there are at least two literally concrete possible worlds.

(F) is clearly entailed by (D) and (E); if an entity exists in one world but not in another, the two worlds cannot be numerically identical.

11. Chalmers (1996, 96) has an interesting discussion of this issue, holding that the burden of proof is on those who claim that something is logically impossible:

> In general, a certain burden of proof lies on those who claim that a given description is logically *impossible*. If someone truly believes that a mile-high unicycle is logically impossible, she must give us some idea of where a contradiction lies, whether explicit or implicit. . . . If no reasonable analysis of the terms in question points to a contradiction, or even makes the existence of a contradiction plausible, then there is a natural assumption in favor of logically possibility.

12. This follows from "$p \rightarrow \Diamond p$," a fundamental and noncontroversial claim in modal system T.

13. I have in mind certain claims that Spinoza makes in Part One of the *Ethics*; e.g., he states that if something possibly exists in the actual world (or if there is nothing that prevents it from existing in the actual world), then it exists in the actual world. Of course, an unpopular philosophical claim might still be true; nevertheless, many philosophers will be reticent to deny (2) given what a denial (2) entails.

14. In (A), a conditional, the antecedent is equivalent to the consequent, except for one crucial difference: in the antecedent, *e* is a concrete entity that could be added to the actual world without generating a contradiction, while in the consequent, *e* is possibly concrete in the actual world. That is, we move from the absence of contradiction in the antecedent to logical possibility in the consequent. The point is that if indeed the lack of a contradiction entails logical possibility, then (A) is true.

15. These are the only two possibilities. Either the presence of an entity would generate a contradiction in and of itself (because it is self-contradictory), or not (this is an instance of excluded middle, and so is a logical truth). But if not, then it must generate a contradiction when combined with things outside of itself.

16. The argument for (B) claims that no contradiction would be generated by adding a duplicate of an entity to the actual world, and so, given (A), two numerically distinct entities *could be* (or possibly are) qualitatively identical. However, the Principle of the Identity of Indiscernibles, "first explicitly formulated by Wilhelm Gottfried Leibniz in his *Discourse on Metaphysics*, Section 9 (Loemker 1969: 308) . . . states that no two distinct things exactly resemble each other" (Forrest 2010, Introduction). So if this argument is successful, Leibniz's principle cannot be necessarily true, at least.

17. This argument, unlike the previous (second) argument for (2), does not depend upon the claim that consistency entails logical possibility.

18. So this argument also conflicts with Leibniz's Principle of the Identity of Indiscernibles. One might object that it is not obvious that one could simply add, say, another grain of sand to the actual world without producing an impossibility, or at least massive changes in the actual world. For example, one might claim that the laws of nature are so fine-tuned and interconnected that adding a grain of sand to the actual world would require massive changes. But the laws of nature are not so fine-tuned that they entail that everything that exists in the actual world must exist and that nothing that doesn't exist in the actual world couldn't. The laws of nature could be held constant while the number of things in existence changes. Adding another grain of sand to the actual world would require there to be more matter in the world; but the

laws of nature seem consistent with that possibility; there would simply be a little bit more matter following the same laws as all the other matter, etc.

19. To clarify, (A) is essentially a universally quantified claim; "for all p, if p is logically possible, then there is a logically possible world w in which p obtains". So (B) can be inferred from (A) with universal elimination; simply substitute "collection of facts" for p.

20. Imagine if someone said "there is a logically possible world that contains only cats," but then denied that there is a logically possible world that contains only cats. So there both is and is not a logically possible world that contains only cats. Indeed, a denial of (A) generates a contradiction.

21. Note that this is not just the definition of "possibility" in possible worlds semantics; indeed, (A) is more trivial than that.

22. Note that these propositions will be true in the relevant world and so will have truth-values in general; see, for instance, Parent (2012, Section 3): "After all . . . the Ersatzers generally speak of what is true 'according to a world'". Indeed, these propositions that make up worlds must have truth-values; a world will be a set of propositions, and some propositions will be in a given set that is a world while others will not. And it must be the truth-value of the proposition in that world that determines if the proposition is in the set or not; if a proposition is true in a world, the proposition will be in the set, but it will not be in the set if it is false in that world (it would be incoherent to claim, e.g., that a given car is possibly red in a world, and that this is captured by the fact that the proposition "the car is red" is in the set of propositions that compose a world, yet that proposition is false in that world, etc.). Moreover, we are told that these sets of propositions are consistent; so it is never the case that a given proposition is both true and false in a world, but then they must have truth-values. Talk of propositions being consistent (or inconsistent) with one another makes no sense if the propositions in question lack truth-values.

23. Note that ersatzism is consistent with the possibility of a multiverse if (ii), i.e., if a possible world can contain a multiverse. For then it might be that there is only one concrete possible world (ours) and it contains a multiverse.

24. Note that if the argument for restricted modal realism is sound, then we know that there are at least two CONCRETE possible worlds. So if (i) a possible world can contain only one universe, then the universes that are CONCRETE are spread out across different possible worlds and each world will contain— or will be—one universe. But if (ii) a possible world can contain more than one universe, there might be multiverses in multiple possible worlds.

25. See also Ratzsch (2010, Section 4.2):

> The traditional method of overcoming prohibitive single-throw odds has been to multiply the number of tries—much as one can overcome the odds against throwing a double six given enough throws of the dice. In general, a state space of possibilities, no matter how extensive, can be saturated via enough separate random tries, so that any value-points in the space will eventually be discovered. Hume discussed this type of strategy for countering cosmic design arguments, and current many-universe theories are sometimes intended to function in similar manner, thus undercutting cosmic fine-tuning arguments.

26. I would like to thank Michael Schrynemakers, Klaas Kraay, and the participants at the God and the Multiverse Workshop (Ryerson University, February 15–16, 2013) for immensely helpful comments on an earlier draft.

REFERENCES

Almeida, M. 2008. *The Metaphysics of Perfect Beings*. New York: Routledge.
Carnap, R. 1947. *Meaning and Necessity*. Chicago: University of Chicago Press.
Chalmers, D. J. 1996. *The Conscious Mind: In Search of a Fundamental Theory*. New York: Oxford University Press.
Collins, R. 1999. "A Scientific Argument for the Existence of God: The Fine-Tuning Argument." In *Reason for the Hope Within*, edited by M. J. Murray, 47–75. Grand Rapids, MI: Eerdmans.
Daly, C. J. 2008. "The Methodology of Genuine Modal Realism." *Synthese* 162: 37–52.
Forrest, P. 1981. "The Problem of Evil: Two Neglected Defences." *Sophia* 20: 49–54.
Forrest, P. 2010. "The Identity of Indiscernibles." In *The Stanford Encyclopedia of Philosophy*, edited by E. N. Zalta. http://plato.stanford.edu/entries/identity-indiscernible/.
Himma, K. E. 2006. "Design Arguments for the Existence of God." In *Internet Encyclopedia of Philosophy*, edited by J. Feiser and B. Dowden. http://www.iep. utm.edu/d/design.htm.
Jackson, F., G. Priest, and K. McDaniel. 2004. "Modal Realism with Overlap." *Australasian Journal of Philosophy* 82: 137–52.
Kraay, K. 2010. "Theism, Possible Worlds, and the Multiverse." *Philosophical Studies* 147: 355–68.
Kripke, S. 1963. "Semantical Considerations on Modal Logic." *Acta Philosophica Fennica* 16: 83–94.
Lewis, D. 1986. *On the Plurality of Worlds*. Oxford: Blackwell.
Loemker, L., ed. and trans. 1969. *G. W. Leibniz: Philosophical Papers and Letters*. 2nd ed. Dordrecht: D. Reidel.
McHarry, J. D. 1978. "A Theodicy." *Analysis* 38: 132–134.
Megill, J. 2011. "Evil and the Many-Universes Response." *International Journal of Philosophy of Religion* 70: 127–38.
Miller, R. B. 1993. "Genuine Modal Realism: Still the Only Non-Circular Game in Town." *Australasian Journal of Philosophy* 71: 159–60.
Miller, R. B. 2001. "Moderate Modal Realism." *Philosophia* 28: 3–38.
Parent, T. 2012. "Modal Metaphysics." In *Internet Encyclopedia of Philosophy*, edited by J. Feiser and B. Dowden. http://www.iep.utm.edu/mod-meta/.
Plantinga, A. 2003. *Essays on the Metaphysics of Modality*. Edited by M. Davidson. Oxford: Oxford University Press.
Ratzsch, D. 2010. "Teleological Arguments for God's Existence." In *Stanford Encyclopedia of Philosophy*, edited by E. N. Zalta. http://plato.stanford.edu/entries/ teleological-arguments/.
Rosen, G. 2012. "Abstract Objects." In *The Stanford Encyclopedia of Philosophy*, edited by E. N. Zalta. http://plato.stanford.edu/entries/abstract-objects/.
Spinoza, B. (1677) 1985. *The Collected Works of Spinoza*. Vol. 1. Edited and translated by E. Curley. Princeton, NJ: Princeton University Press.
Szabo-Gendler, T., and J. Hawthorne, eds. 2002. *Conceivability and Possibility*. Oxford: Oxford University Press.
Tegmark, M. 2003. "Parallel Universes." In *Science and Ultimate Reality: From Quantum to Cosmos*, edited by J. D. Barrow, P. C. W. Davies, and C. L. Harper, 459–91. Cambridge: Cambridge University Press.
Turner, D. 2003. "The Many-Universes Solution to the Problem of Evil." In *The Existence of God*, edited by R. Gale and A. Pruss, 143–59. Aldershot, UK: Ashgate.

5 Revisiting the Many-Universes Solution to the Problem of Evil

Donald A. Turner

In an earlier essay, I defined a 'simple possible world' to be one that describes a single maximal spatiotemporal aggregate, a cosmos or universe, and a 'complex possible world' to be one with multiple cosmoi or universes (Turner 2003). The instantiation of a simple possible world will be a single universe, and I called the instantiation of a complex possible world a 'multiverse'. I continued,

> If a wholly good and omnipotent God exists, then the fact that it would be best if created reality were a certain way does explain why created reality would be that way. The source of selection from among possible universe ensembles would be the possible universe ensemble that would be best. Thus I claim that God ought to actualize that complex possible world which contains cosmoi [or universes] corresponding to every simple possible world above some cut-off line—for example, every simple possible world with a favorable balance of good over evil. (149)

There have been a number of responses to multiverse solutions to the problem of evil. In this essay, I consider and respond to objections due to Bradley Monton (2010), Michael Almeida (2008), and Klaas Kraay (2012), in turn.

MONTON

Monton (2010) rejects the Principle of the Identity of Indiscernibles. While I still think that the Principle of the Identity of Indiscernibles is true, suppose that I am wrong about that, and that God can create duplicates of universes. (On my view, duplicates of universes would be identical because all of their intrinsic properties would be the same, and since universes bear no spatiotemporal relations to one another, I don't see how they could have differing relational properties. It is not as though one universe could bear one spatiotemporal relation to ours, and its duplicate a different spatiotemporal relation to ours, since neither has any spatiotemporal relation to ours, to one another, or to any other universe at all. Nor do I see any other relational properties by which they could differ.) While I don't think that God can do the logically impossible, I see no reason to think that God would be limited to any given infinite cardinal in his duplication. I see no reason to think that for any number, no matter how large an infinite cardinal that number was, that God would be limited to that number. The idea that God has to pick a number of universes to create, whether that number is finite or infinite, seems to me to be an unjustified limitation on his power. That is, I think that God would create all possible duplicates of every universe with a favorable balance of good over evil, so that no matter what infinite cardinal you consider, whether 2^n, $2^{(2^n)}$, $2^{[2^{(2^n)}]}$, $2^{\{2^{[2^{(2^n)}]}\}}$, and so on, God will have created more than that number of duplicates. God will have created the best possible world, one which contains all possible duplicates of every universe with a favorable balance of good over evil.

Monton (2010, 130) continues:

> Imagine that, in the collection of universes God creates, there is a universe *U*. This universe has a favorable balance of good over evil, but it contains a sentient creature who seemingly pointlessly suffers. Specifically, there is no reason, based on just what happens in *U*, for that creature to suffer—there is no countervailing good that comes of it, for example. Imagine that there is also, in the collection of universes God creates, a universe I'll call *Nice-U*, which is qualitatively identical to *U*, except that the counterpart of the pointlessly suffering creature does not undergo the pointless suffering. (For example, the creature might become a zombie for the time period its counterpart pointlessly suffers, or the creature might simply cease to exist for that time period.) What is the virtue of creating *U* in addition to *Nice-U?* Instead of creating *U*, why didn't God just create a duplicate (or near-duplicate) of *Nice-U?* I maintain that there is no reason for God to create a universe where creatures pointlessly suffer; he is better off creating duplicates of nice versions of those universes instead, where no pointless suffering happens.

Suppose that the Principle of the Identity of Indiscernibles is false. Then, along the lines of my preceding comments, it seems to me that God would get the best overall creation by creating all of the possible duplicates (more than any infinite cardinal) of Nice-U *and* all of the possible duplicates of U. While one can't simply sum up and compare utilities (as the utilities of all possible duplicates of Nice-U, and the utilities of all possible duplicates of Nice-U together with all possible duplicates of U, will both transcend any transfinite number) each possible duplicate of U adds utility that would not exist if God were not to create that duplicate. This is the same basic reason why there is value in God creating anything at all outside himself when he alone is already of infinite worth.

ALMEIDA

Michael J. Almeida (2008, 145) claims:

> The existence of individuals across countless worlds makes at least some multiverses impossible. There is a world w in which Socrates is married to Xantippe and a world w' in which Socrates is not married to Xantippe. But there could be no multiverse including both w and w'. And generally there are no multiverses that locate individual persons, particles, leptons, waves, or spirits in multiple universes.

On my view, there *is* a complex possible world including both w and w'. Call the Socrates in w' Socrates'. Obviously Socrates and Socrates' are not numerically identical (see later). But note that this is true for the modal realism that Almeida is exploring as well.

According to Mark Heller (2003, 1),

> Modal realists . . . believe that the actual world is a concrete object of which you and I are literal parts, and that other worlds are also concrete objects some of which literally include other people as parts. Merely possible worlds and merely possible people *really exist* despite their lack of actuality.

On this view, all possible worlds are on an ontological par, and 'actual' is merely an indexical term picking out the world that one is in just as 'here' picks out the place that one is in. This world is not more real than other possible worlds any more than this place is more real than other places. For modal realists, there is a world w in which Socrates is married to Xantippe and a world w' in which Socrates' is not married to Xantippe'. Socrates and Socrates' are not numerically identical. Instead, modal realists hold that Socrates' is a counterpart of Socrates. I can help myself to this exact same counterpart relationship.

Now some people do not think that the counterpart relationship is enough. On my view, the relationship between Socrates and Socrates' is actually very like that between identical twins. While Socrates' is not Socrates, I would say that Socrates' is someone whom Socrates could have been; indeed he is who Socrates would have been if Socrates had had the experiences that Socrates' had, that is, if the boy who grew up to become Socrates had had the experiences that Socrates' had instead of those that Socrates had, those experiences would have shaped him in a different direction, so that he would have become Socrates'. I would claim the same regarding other individual persons, particles, waves, and so forth. If w' differs from w only in ways that do not affect Socrates, then I would say that while Socrates' is not numerically identical to Socrates, he does bear the relationship that we call 'personal identity' to Socrates. I deny that personal identity is identity, strictly speaking; for example, I do not think that personal identity is transitive. (Note that on my view, God is outside of all worlds, so there is no question of whether God in one world is identical to God in another.)

Almeida continues: "Turner's proposal that there actually exists a complex world seems to entail that any number of logically inconsistent propositions are all actually true. That of course is *prima facie* impossible" (Almeida 2008, 147). Almeida correctly anticipates my response when he says, "He seems to have in mind that truth in the actual world is better understood as *truth-at-a-cosmos* in the actual world" (147). This is correct. In the same way that there is no contradiction between saying that it is snowing here and now, but that it is not snowing at certain other times and places, there is no contradiction between saying that Socrates is married to Xantippe in w and Socrates is not married to Xantippe in w'. This means that it is true both that Socrates is married to Xantippe and that Socrates is not married to Xantippe, but this is no more of a contradiction than that it is true both that it is snowing and that it is not snowing. Both claims need to be made more precise: it is snowing here and now, and it is not snowing at other times and places. Socrates is married to Xantippe in w, and Socrates is not married to Xantippe in w', though strictly speaking, of course, "Socrates is not married to Xantippe in w'" means "Socrates' is not married to Xantippe' in w'".

Almeida continues:

> What Turner is suggesting, presumably, is that there is a complex world W that includes many spatiotemporally isolated simple worlds w. Indeed, there is alleged to be a complex world W that includes *every* on-balance good, simple world w. But how is this possible? No simple world w can be both included in complex world W and not included in complex world W. Turner might be supposing that there's some complex world W that includes every simple on-balance good world w. This entails that there are no on-balance [good] simple worlds that are not included in a multiverse W. Call that the *Wonderful Complex World Assumption* . . . The *Wonderful Complex World* assumption seems implausible. Why believe that every simple on-balance good world is included in some single multiverse W? Why not believe instead that the simple on-balance good worlds are spread out among many multiverses or included in no multiverses at all? Obviously these cannot all be true. (148–49)

To make things simple, suppose for the moment that there are only four simple possible worlds, two that are on-balance good, and two that are on-balance bad. Call the on-balance good simple possible worlds w and w' and the on-balance bad possible worlds w'' and w'''. Then there will be 11 complex possible worlds, the one containing simple possible worlds w, w', w'', and w''', the one containing w, w'', and w''', the one containing, w, w', and w''', the one containing w, w', and w'', the one containing w', w'', and w''', the one containing w and w', the one containing w and w'', the one containing w and w''', the one containing w' and w'', the one containing w' and w''', and the one containing w'' and w'''. There will be a multiverse if and only if God actualizes one of these complex possible worlds. Each simple possible world is a part of many complex possible worlds. God would also have the option of actualizing any one of the simple possible worlds alone by creating only the single universe instantiating it, or of actualizing the empty world by creating nothing at all. There will be a complex possible world containing only the simple on-balance good possible worlds because there will be a complex possible world containing *every* possible combination of simple possible worlds. But my argument is that a perfectly good God should actualize that complex possible world that contains only the on-balance good simple possible worlds, in this example, that complex possible world containing simple possible worlds w and w'.

Almeida continues:

> Let U be the set of all on-balance good universes and all on-balance neutral universes . . . Suppose a perfect being actualizes a multiverse W that includes every member of U. The standard claim is that W is the best possible world . . . But it is impossible that any universes in U should form the spatiotemporally isolated island universes of W unless no two universes in U overlap with respect to persons, events, objects, or states of affairs, among many other things. . . But there is very good reason to believe that there is overlap among the universes that are on-balance good. Suppose the actual universe is an on-balance good universe in U. Now consider whether there is a universe u' exactly like u except that some rational and sentient beings in u are slightly better off in u'. It is difficult to deny that some actual rational and sentient beings might have been slightly better off. But then, if we take this conclusion at face value, there are at least members of U that overlap with respect to some rational and sentient beings. Universes u and u' are therefore not compossible. (149–50)

It should be clear at this point what my response is. To say that Socrates might have been better off than he is in universe u is to say that there is someone whom Socrates could have been, or if one prefers, a counterpart of Socrates, Socrates' and a universe u', such that Socrates' is better off in u' than Socrates is in u, the very same answer that the modal realist gives, and thus universes u and u' *are* compossible.

Almeida continues:

> In some worlds that perfect being has the property of always acting providentially. In other worlds that perfect being has the property of rarely acting providentially. But there is obviously no world that includes an island universe u where a perfect being has the property of

always acting providentially and another island universe u′ in which the *very same* perfect being has the property of rarely acting providentially. So those universes are not compossible. One resolution of these difficulties might urge that we index the properties of perfect beings and properties to worlds. We should then identify the property of always acting providentially with the property of always acting providentially in *w* and the property of rarely acting providentially in *w*′. (150)

If, despite the perfect being rarely acting providentially in *w*′, *w*′ is nevertheless one of the simple possible worlds above the cutoff line, then this is just the response that I would give. This alone makes God always acting providentially necessary no more than its snowing here and now is a necessary truth simply because it is true in all possible worlds that it is snowing here and now in this world.

Almeida continues:

> But it is evident that these properties are not identical. They are not even mutually entailing. The property of always acting providentially, for instance, is not identical to the property of always acting providentially in *w*. The latter is an essential property that perfect beings instantiate in every possible world. The former is a contingent property that perfect beings instantiate in some worlds but not others. Even in worlds where a perfect being rarely acts providentially, he instantiates the property of always acting providentially in *w*. (150–51)

All of this is true, but I don't see why it is a problem. Suppose that God causes manna to fall in Toronto on February 17, 2013. Now compare the property of causing manna to fall and causing manna to fall on February 17, 2013. These properties are not identical. The former does not entail the latter. The latter is one that God instantiates at all times and places. But that is just what one would expect, and the case discussed earlier seems to me to be just like this one. I don't see why it is supposed to be a problem. Just as "it is snowing" is most naturally taken to mean "it is snowing here and now", it seems to me that "God always acts providentially" is most naturally taken to mean "God always acts providentially in this universe".

Recall that on my view, God is actually outside of all possible worlds. Just as a defender of divine eternity would say that God is outside of time, and that God's becoming incarnate and God's causing manna to fall on February 17, 2013 are both part of the single divine action, I would say that God always acting providentially in *w* and God's rarely acting providentially in *w*′ are part of a single divine action.

Almeida continues:

> Consider any cosmos u in the actual multiverse *W* that contains total good uG and total evil uE. Since u is a member of the actual multiverse,

we know that (uG—uE) > 0 or that u is on balance good. Now consider some amount of evil uE' = 1/2 (uG—uE) that is causally and logically independent of every event, object, and state of affairs in u. The universe u' that results from adding the evil uE' to the universe u is also a member of the actual multiverse. We know that u' is also on balance good since the overall value of u' = (uG = (uE + uE')) > 0. The universe u' is on balance good and every universe that is on balance good is a member of the actual multiverse.

So there is a member of the actual multiverse that contains at least some evil uE' that is *causally and logically independent* of every event, object, and state of affairs in u'. But then the evil uE' obviously serves no purpose in u'. It is perfectly possible to remove uE' from u' without the loss of any positive value in u'.

Since a perfect being might have prevented uE' in u' without any moral cost, the standard account of gratuitous evil would categorize uE' as an instance of gratuitous evil. But *Turner's Multiverse Solution* entails that uE' is not an instance of gratuitous evil . . . Suppose uE' is William Rowe's well-known isolated and painful death of a fawn. The isolated and painful death is justified, according to *Turner's Multiverse Solution,* if uE' occurs in some world that is on balance good. A perfect being could not prevent the isolated and painful death of the fawn without failing to actualize a world that is on balance good. (151–52)

In one sense, Almeida is of course correct when he points out that it is perfectly possible to remove uE' from u' without the loss of any positive value in u'. But it is not correct that a perfect being might have prevented uE' in u' without any moral cost. If the perfect being had prevented uE' in u', then u' wouldn't exist as a separate universe at all, and so all of the goodness in u' would have been lost. It is true that on my view, what on the standard view of gratuitous evil is an instance of gratuitous evil turns out not to be gratuitous after all. But, as the saying goes, "that's not a bug; it's a feature". It is difficult to reconcile even one instance of genuinely gratuitous evil with the existence of a perfect being. The fact that on my view there is no genuinely gratuitous evil seems to me to be a virtue of my view.

Almeida continues:

Turner's Multiverse Solution in fact entails that every instance of actual evil is necessary to the actualization of the best possible multiverse. A perfect being is therefore justified in permitting every instance of actual evil. But the multiverse solution to the problem of evil generalizes in very problematic ways . . . Of course the argument generalizes to every instance of actual evil, and every actual evil is therefore justified evil. But the argument also generalizes to every event, object, and state of affairs that occurs, exists, and obtains in the actual multiverse.

Turner's Multiverse Solution entails *Necessitarianism*. Necessitarianism is the (false) position that there is exactly one possible world. (152)

For reasons I will explain presently, there is more than one possible world on my view. But it is true that, in one sense, there are significantly fewer than it might first appear. But presumably this is true at least to some extent for anyone who believes that this universe was created by an essentially good God. Consider a world in which there is much more evil than good. Is this world possible? Presumably very few who believe in an essentially good God would believe that God would have actualized such a world. So is that world possible? It depends on what one means by 'possible'. In one sense, that world is possible. In another sense, that world is not. To use an analogy, suppose that Gandhi is in a room with an AK-47, and someone is walking by outside the window. Does Gandhi have the power to shoot the person? Well, it depends on what one means by 'power'. In one sense, he has a gun which has the power to shoot the person. But his character is such that he would not use that power. Similarly, the world in which there is much more evil than good is possible in the sense of being *internally* consistent. As I said previously, on my view, God exists outside of all possible worlds, surveying them and deciding which to actualize, rather than in any of them, and so the internal consistency of a world is determined without reference to whether it is a world the actualization of which is consistent with God's character. Call this 'possible$_1$'. But that world is not possible in the sense of being a world that an essentially good God would ever actualize, just as Gandhi would never use the power of the AK-47 to shoot the person walking by. Call that 'possible$_2$'. On my view, there are as many worlds that are possible$_1$ as on anyone's view; perhaps more, since on my view there are the complex possible worlds as well as the simple ones. But it is true that there are significantly fewer worlds that are possible$_2$. Indeed, among the possible$_1$ worlds that were available to God to actualize, there is indeed only one that is possible$_2$. But I think that there are worlds that are possible$_2$ that were not available for God to actualize. On the other hand, it seems to me that the view Almeida explores *is* necessitarian, for while I realize that a modal realist wouldn't describe it this way, modal realism seems to me to be much like saying that necessarily that complex possible world exists which contains universes corresponding to *every* simple possible world.

Almeida continues:

> Of course the necessitarian conclusion is wildly implausible. In necessitarian worlds there is obviously no free will and no moral responsibility. In necessitarian worlds there is no agency, no basis for self-respect, or moral praise or blame. (153)

I accept Molinism, the view that there are true counterfactuals of creaturely freedom which specify what God's free creatures would in fact do in any situations that they were free with respect to. Possible worlds in which the counterfactuals of creaturely freedom are different from what they in fact are were not available for God to actualize; they are not feasible. If God creates beings with genuine libertarian free will, then to some extent, creation becomes a cooperative venture among God and all of his free creatures, and to get the best possible world, not only must God do his part, but his free creatures must do theirs as well. But if they are genuinely free, then whether they do their part is up to them, not up to God. So among the infeasible worlds, the ones with different counterfactuals of freedom, there are many that are possible$_2$, ones such that God's actualizing them would have been consistent with his goodness, ones that he would indeed have actualized had they been feasible. So strictly speaking, my view is not that God actualizes the best possible world, but rather the best possible world that he can, the best feasible world. This will still be a multiverse except in the extremely unlikely case that there is only one simple feasible world with a favorable balance of good over evil. And on this view, the actual complex possible world that God creates will not be necessary. There will be other worlds that are possible$_2$. If we, his free creatures, were to make different free decisions from the ones which we would actually make, which we are indeed free to do, then God would actualize a different complex possible world instead. Though, given that on my view, God is essentially such that he always acts for the best, it would of course be the best possible world that he could actualize given *those* free decisions.

Almeida argues that my view, along with the hyperspace view of Hud Hudson (2005), entails that there is no divine freedom (162–64). On this count, I must plead, "guilty as charged". But again, this is not unique to Hudson's and my views. Essentially perfect goodness and morally significant free will are incompatible properties. If a being has morally significant free will, then it is possible for that being to act immorally, or at least to make a moral choice which is less than the best; and if a being is essentially perfectly good, then it is impossible for that being to make a moral choice which is less than the best. Thus anyone who thinks, as I do, that God is essentially perfectly good, has to deny that God has morally significant free will. It is true that my view holds that there is no divine freedom at all, whereas someone could hold while God does not have morally significant free will, he is free with respect to certain morally insignificant choices, but I don't see anything to be gained by holding that God has morally insignificant free will. Another reason why divine freedom is sometimes thought to be necessary is the fear that without divine freedom, the favor that God displays toward us would not be gracious. But this is not true, because God does not owe it to *us* to treat us well; rather, he owes it to *himself* always to act for the best. My view does entail that God is not morally praiseworthy in the technical sense of being morally responsible for a morally good choice, but it does not follow from that that God is not praiseworthy for other reasons, including for his essential goodness.

Indeed, I would claim that his essential goodness is reason for the greatest praise of all, and thus I conclude that God is indeed a proper object of worship. One might wonder, if moral praiseworthiness is not necessary in God's case, why would he create *us* with morally significant freedom? The reason is because *both* essential goodness and moral praiseworthiness are great-making properties, and while they cannot be combined in the very same being because they are not compossible, the best possible world will contain beings with each.

KRAAY

Klaas J. Kraay, in response to my suggestion (and similar suggestions by others) that God should actualize that complex possible world containing all (and as he correctly points out) only universes corresponding to every simple possible world above a certain cutoff level, for example, every simple possible world with a favorable balance of good over evil, says (Kraay 2012, 148):

> But such a threshold is too simplistic: a universe might meet this condition, while nevertheless containing some feature that makes it unworthy of inclusion in a divinely furnished multiverse. Perhaps sensing this, some philosophers have proposed additional requirements. Parfit says there should be no injustice (1991, p.5), and that each individual's life must be worth living (1992, p.423). Forrest first says that every individual must have a life in which good outweighs evil (1981, p.53), and later adds two further restrictions: each creature who suffers must at least virtually consent to it, and must receive ample recompense afterwards (1996, pp.225–7). Draper says that no individual's life may be bad overall, and that God must be a benefactor to all creatures (pp.319–320). These criteria may well be plausible, but of course they are only partial specifications of a threshold, since they only pertain to universes inhabited by creatures like ourselves. A more complete account of the threshold is needed.

Parfit's first requirement, that there should be no injustice, seems to me to be too strong. God's essential goodness would not allow him to commit injustice, but I do not see that it would prevent him from *permitting* injustice so long as some greater good came out of that injustice. So far as Parfit's second requirement, that each individual's life must be worth living, I would say that God's goodness would require that if any individual's life was not worth living, that the individual would have be morally responsible for the fact that his or her life was not worth living. But in considering whether an individual's life was worth living, we would have to take into account, not just this life, but the afterlife as well. I would give this response as well to Draper's restrictions as well as to Forrest's initial restriction, that every individual must have a life in which good outweighs evil, as well as to the second of his later restrictions, that each creature who suffers must receive ample recompense afterwards. In response to the suggestion that we would have to take into account, not just this life, but the afterlife as well, Draper says that God would have a *prima facie* moral obligation to immediately remove creatures from earth, to prevent them from having a bad earthly life, and that God would have this obligation "even if that would lower the overall value of God's creation taken as a whole" (Draper 2004, 319–20). My intuitions differ from his. I simply see no reason to think that this is true. So far as the first of Forrest's later restrictions, that each creature who suffers must at least virtually consent to it, I simply fail to share his intuition on that point as well. On Kraay's point that a more complete account of the threshold is needed, I agree, though I doubt that a complete account of the threshold is possible. One's view of the threshold will depend in part on one's ethical theory.

Kraay concludes his essay by saying, "Theistic multiverse theories face several important problems, but also offer significant prospects for various issues in the philosophy of religion. Solving the problem of evil, however, is not among these prospects" (Kraay 2012, 158). That this is not a complete solution to the problem of evil, I certain agree. Whatever the cutoff level, the threshold turns out to be, it is not obvious that our universe is above that threshold, but given that it is certainly more plausible to think that our universe is above that threshold than to think that it is the instantiation of the single best simple possible world, if such there be, I do think that this argument shows it to be more plausible than some have thought that this universe is one that a wholly good God would have created.[1]

NOTE

1. I thank Michael Almeida, Klaas J. Kraay, and the participants in the God and the Multiverse Workshop (Ryerson University, February 15–16, 2013) for their helpful comments on an earlier draft of this essay.

REFERENCES

Almeida, M. 2008. *The Metaphysics of Perfect Beings.* New York: Routledge.
Draper, P. 2004. "Cosmic Fine-Tuning and Terrestrial Suffering: Parallel Problems for Naturalism and Theism." *American Philosophical Quarterly* 41: 311–21.
Forrest, P. 1981. "The Problem of Evil: Two Neglected Defenses." *Sophia* 20: 49–54.
Heller, M. 2003. "The Immorality of Modal Realism, Or: How I Learned to Stop Worrying and Let the Children Drown." *Philosophical Studies* 114: 1–22.
Hudson, H. 2005. *The Metaphysics of Hyperspace.* Oxford: Oxford University Press.

Kraay, K. J. 2012. "The Theistic Multiverse: Problems and Prospects." In *Scientific Approaches to the Philosophy of Religion*, edited by Y. Nagasawa, 143–62. Houndsmill, UK: Palgrave Macmillan.

Monton, B. J. 2010. "Against Multiverse Theories." *Philo* 13: 113–35.

Parfit, D. 1991. "Why Does the Universe Exist?" *Harvard Review of Philosophy*, Spring, 4–5.

Turner, D. 2003. "The Many-Universes Solution to the Problem of Evil." In *The Existence of God*, edited by R. Gale and A. Pruss, 143–59. Aldershot, UK: Ashgate.

Criticisms of Theistic Multiverses

6   Kraay's Theistic Multiverse

Michael Schrynemakers

Klaas Kraay (2010) argues that theists have good reason to think that we live in a multiverse comprising all and only universes worthy of being created and sustained by God. He calls the world in which God actualizes this multiverse the 'Theistic Multiverse' (TM) and argues that TM is the unique greatest possible world. TM is an elegantly simple and helpful model. Kraay argues that this model refutes *a priori* atheistic arguments based on the claim that there is No Unsurpassable World and those based on the claim that there are Multiple Unsurpassable Worlds. He also argues that although TM narrows the evidential basis for *a posteriori* atheistic arguments from the apparent Surpassability of the Actual World, TM does not help solve the problem of evil, contrary to what other philosophers have claimed. After a brief clarification of terminology, I will examine Kraay's TM and how it relates to these arguments. I will argue that this relation is more complex than it first appears and I will disagree with Kraay about how TM bears on these three arguments for atheism.

KRAAY'S THEISTIC MULTIVERSE

A possible world is a perfectly specific possible way that all of reality could be. Since reality may consist of more than one universe, possible worlds and possible universes are importantly distinct. By *universe*, Kraay (2010, 357) means a "spatiotemporally interrelated, causally closed aggregate". This refers to a maximally specific spatiotemporally interrelated history and not only a type of universe. God *actualizes* a world when God brings it about that reality is one way rather than another. As Kraay points out, God's actualizing a world or creating a universe does not necessarily mean that God determines every feature of that world or universe. Human choices and natural indeterminism may contribute to the world's being the way it is, in addition to God's actions. As Kraay also notes, God's actualizing a world may be thought of as consisting of a single act or as many actions at different times. We should also keep in mind that, given theism, God exists in every possible world because if God exists, He exists necessarily, not contingently. Although Kraay does not explicitly mention this, a world actualized by God may be thought to contain immaterial created beings as well as material entities, so I will consider Kraay's TM to include all and only those immaterial entities worth creating (if such there be) as well as material universes.

Kraay's TM, then, is a world consisting of God, anything that exists necessarily (perhaps including uncreated abstract objects that depend on the mind of God), all entities, including universes, worthy of being created and sustained, and no entities, including universes, that are unworthy of being created and sustained.[1] What does it mean for a universe to be worthy of being created and sustained by God? This is a pivotal question, as we shall see. A universe's containing more good than evil is a plausible necessary condition, but Kraay is careful to avoid assuming that this is sufficient. Perhaps a universe is worthy only if it does not exhibit certain bad-making properties, such as containing gratuitous evil or involving God's violation of deontological constraints. Perhaps exhibiting such bad-making properties would also add sufficient disvalue to the universe such that a universe's being net good remains the sole criterion for creation worthiness, but Kraay wisely does not assume this in his argument.

KRAAY'S ARGUMENT THAT THE THEISTIC MULTIVERSE IS THE GREATEST POSSIBLE WORLD

Kraay argues that TM is the greatest possible world. There is no reason to think that God cannot create multiple universes and it seems that the more creationworthy universes He creates, the better. God's creating any unworthy universe would make creation overall worse. So the best possible world includes *all and only* those universes worthy of being created and sustained (TM). TM is the unique unsurpassable possible world, according to Kraay, because any world distinct from TM must either fail to include a worthy universe or include an unworthy universe, and in either case it would be less good than TM. Since God would actualize the greatest possible world, theists should think TM is the actual world.

Kraay articulates this reasoning in terms of a principle of plenitude and principles of restricted creation and sustenance. The principle of plenitude and its sustenance addendum are:

PP1 If a universe is *creatable* by an unsurpassable being, and *worth creating* (i.e., it has an axiological status that surpasses some objective threshold t), that being will create that universe.

PP3 If a universe is *sustainable* by an unsurpassable being, and *worth sustaining* (i.e., it has an axiological status that surpasses some objective threshold t), that being will sustain that universe. (361–62)

The principles of restricted creation and sustenance are:

PR1 If a universe is *not worthy of creation* (i.e., it has an axiological status that fails to surpass some threshold *t*), an unsurpassable being
will not create that universe.

PR3 If a universe is *not worthy of being sustained* (i.e., it has an axiological status that fails to surpass some threshold *t*), an unsurpassable
being will not sustain that universe. (362)

According to Kraay, these four statements imply that God would create and
sustain all and only possible universes worthy of actualization, and thereby
actualize the greatest possible world, TM (362–63).

Kraay's claim that TM is the unique greatest possible world requires
clarification. One might object that, if there are free creaturely choices
or indeterministic processes whose outcomes are not determined by God,
then TM, a multiverse comprising all and only worthy universes *God can
create and sustain,* may not be the greatest possible world. For example,
even if the actual world is TM, it seems plausible to suppose that human
choices could have been better, and that, accordingly, so could TM. In
response, one might deny that the possibility of better free choices and
better indeterministic outcomes in worthy universes means that TM could
have been better, by taking TM to include all choice possibilities and all
indeterministic possibilities of all worthy universes, provided that they do
not result in un worthy universes. (For example, one might hold that if our
universe is included in TM, then TM also includes all possible earth histories no worse than its actual history). But one need not reply to the objection in this way. Kraay's principles refer to universes that are *creatable by
an unsurpassable being,* not to all possible universes, and the arguments
for atheism at issue concern worlds *God can actualize,* not all possible
worlds. Accordingly, throughout this chapter, by 'greatest possible world'
I will mean, as I take Kraay to mean, the greatest possible world, *given free
creaturely choices and results of any indeterministic processes, if these are
not determined by God,* that is, the greatest possible world *an omnipotent
being can actualize.*

THE THEISTIC MULTIVERSE AND
A TRILEMMA FOR THEISM

Kraay next argues that this model defeats two traditional *a priori* atheistic
arguments that involve the claim that there is no unique unsurpassable
world. He then argues that TM shows that atheistic arguments based
on the *a posteriori* claim that *this* is not the unique unsurpassable world
should appeal only to claims about suffering and evil and not to the mere

improvability of our universe. He nicely formulates these traditional challenges to theism as a trilemma:

(a) there is No Unsurpassable World, in which case there cannot be an unsurpassable creator; or
(b) there are Multiple Unsurpassable Worlds, in which case an unsurpassable creator cannot be rational in deciding which world to create; or
(c) there is a unique greatest possible world, in which case God must actualize it, but the actual world is not the greatest possible world. (357)

Here are my standardizations of the arguments suggested by (a)–(c):

Atheistic Argument from No Unsurpassable World

NU1. An omnipotent being could always improve creation by adding some valuable entity or state of affairs.
Therefore,
NU2. For any possible world an omnipotent being actualizes, he could actualize a greater world.[2]
NU3. If an omnipotent being could have actualized a greater world than he did, that being cannot be the greatest possible creator.
Therefore,
NU4. There cannot be a greatest possible creator.

Atheistic Argument from Multiple Unsurpassable Worlds

MU1. Any actualized possible world could have been different in some way that does not affect its axiological status.[3]
Therefore,
MU2. There are at least two unsurpassable worlds God could actualize, if there are any.
Therefore,
MU3. God must either choose a surpassable world, or else must choose one from the set of unsurpassable worlds.
MU4. There is no sufficient reason for an omniscient and perfectly good being to choose one unsurpassable world over another.
Therefore,
MU5. God cannot have a sufficient reason for actualizing one unsurpassable world over another.
MU6. God cannot have a sufficient reason to select a surpassable world rather than a greater world.
MU7. A perfectly rational being must have a sufficient reason for choosing one way rather than another, for every choice he makes.

MU8. If God exists, He is perfectly rational.
Therefore,
MU9. God cannot exist.

Atheistic Argument from the Surpassability of the Actual World

SW1. A perfectly good being would actualize the greatest world he could.
Therefore,
SW2. If God, an omnipotent and perfectly good being, exists, He would actualize the greatest possible world.
SW3. [Probably] God could have either created a better type of initial universe state than ours, or could have improved our universe by preventing certain evils or losses of good.
Therefore,
SW4. [Probably] This is not the greatest possible world.
Therefore,
SW5. [Probably] God does not exist.

Kraay argues that TM refutes the Atheistic Argument from No Unsurpassable World. According to Kraay, TM shows that the second premise, NU2, is false. No world distinct from TM could be greater than TM. For a world to be distinct from TM, it must either not include a universe that is worthy (of being created and sustained), or else include a universe that is unworthy, and either way the resulting world will be worse than TM. So we have good reason to think there is a greatest possible world, according to Kraay; it is TM.

Likewise, Kraay argues that TM refutes the Atheistic Argument from Multiple Unsurpassable Worlds. According to Kraay, TM shows that the first premise, MU1, is false. TM could not have been different without being surpassable. This is for the same reason that TM refutes the Atheistic Argument from No Unsurpassable World: for a world to be distinct from TM, it must either not include a universe that is worthy, or else it must include a universe that is not worthy, and either way the resulting world will be worse than TM. TM is, therefore, the *unique* unsurpassable world, according to Kraay.

The third part of the trilemma for theism, the Atheistic Argument from the Surpassability of the Actual World, affirms that God would actualize the best world within his power to actualize, but claims that the actual world is not that world. Kraay responds that since TM, a multiverse of *all and only* universes worthy of being created and sustained, is the greatest possible world, to judge that the actual world is not the greatest possible world one must judge either that *not all* worthy universes have in fact been created and sustained or that some *un*worthy universes have in fact been created and sustained. One cannot survey all universes to judge whether all worthy possible universes have been created. However, one can claim that *our* universe exhibits properties that show it to be unworthy of being created or sustained. Accordingly, Kraay thinks that although the apparent surpassability of our universe does not by itself indicate this is not the greatest possible world (because, given TM, we have no reason to expect ourselves to be in an unsurpassable universe), he does *not* think that TM, or the more general thesis that we may well live in a multiverse, helps to solve the problem of evil.

Some philosophers have argued that the evil found in our universe cannot be evidence that this is not the greatest possible world because, for all we know, our universe contains more good than evil and, for all we know, the greatest possible world includes *all* universes that contain more good than evil (Turner 2003; Megill 2011). Kraay disagrees. According to Kraay (2012, 2013), since the multiverse God creates and sustains will not contain universes that are unworthy, the multiverse view merely reframes the discussion of the problem of evil, and the appeal to it in the Atheistic Argument from the Surpassability of the Actual World. Considering whether or not our universe is worthy of being created and sustained, as debating the actuality of TM requires, must involve all the questions about how God could be justified in allowing the evils of our universe that are currently under debate apart from consideration of the multiverse view.

AN OBJECTION TO KRAAY'S THEISTIC MULTIVERSE

TM seems to simply and elegantly dissolve two significant objections to theism while supplying needed clarity to the current discussion of the relationship between multiple universes and the problem of evil. In the remainder of this essay I argue that although TM is helpful as a starting point for discussion of the preceding trilemma, it requires reformulation and does not have the implications Kraay claims for it. To begin, consider the following:

> *Alternative Worthy History:* There is at least one worthy universe U that God could have governed differently so as to bring about a different history, such that the result, U′, would also have been a worthy universe.
> *One History Per Universe:* There is no possible world containing both universe U and the 'alternate history universe', U′.

These theses imply that TM is impossible. If God could have actualized a different worthy history than He has, as per the Alternative Worthy History Thesis, then there is a possible universe with that history. But by One History Per Universe, there is no world containing both the actual worthy universe, U, and the possible worthy universe, U′, that would have resulted if He had governed U differently. Therefore, these theses imply that God cannot actualize a world that includes *all possible* worthy universes.

One History Per Universe states that a world cannot include both U and U'. As per Alternate Worthy Universe, U' refers to what U could have been but is not; specifically, U and U' refer to *alternate* ways God may govern a universe. For example, suppose that in U God influences events and people so that David becomes king, but in U', God does not. No possible world can include both U and U' because a universe cannot be governed by God one way and also not governed by God that way. For God to create and sustain U' is for God to choose an alternative to U's actual history. If God could create and sustain both U and U', U and U' would not be divine choice alternatives.

Now, God's creating and sustaining U does not mean that God could not also create and sustain *another* universe qualitatively identical to U'. Again, U' only refers to what U could have been but is not; it does not refer to a type of universe. So, One History Per Universe does not imply there cannot be a universe which is qualitatively just like U in all respects up to the events that led to David's becoming king, but that differs from U in that the person just like David does not become king, for example. But this other universe can only be *just like* the universe that would have resulted if God had governed U differently in this way. It is not U differently governed, which is to say that it is not U'.

To deny One History Per Universe is to deny that God chooses between alternate ways of providentially governing a universe. Can this denial be made plausible? One might think so on the basis of the idea that God's universe-governing activity is not distinct from His universe-creating activity. If God creates and governs a universe in one action, He is not faced with a choice of how to govern a universe He has created, one might think. On this view, there are different possible universes, all of which are compossible, and God's governance consists in His choosing whether or not to create and sustain each one. This objection rejects the presupposition of One History Per Universe that Alternate Worthy Universe must be interpreted as implying that U and U' refer to *alternatives* in the divine choice context. On this view, to say that a universe could have been governed differently by God (so that it could have had a different history) is not strictly speaking correct: it should be translated into the more accurate claim that God could have created and sustained *a* universe with that history.

This objection to One History Per Universe results, however, in a departure from traditional theism. Even if one thinks of God's creating and governing a universe as one action, that action is traditionally thought of as including choices between different alternatives for that universe. If it does not, as this objection suggests, then God never has a choice about *how* a universe history unfolds, or how the lives within it unfold. He only has a choice about whether or not to create distinct universes. Unless individuals can bi-locate, it seems that God cannot choose to influence a person's life in one way *rather than* another, on this view; and if individuals can bi-locate, God need never so choose, on this view. To defend TM in this way is to limit TM to an untraditional form of theism.[4]

What about Alternative Worthy History? This thesis is not as obvious. Let us examine it in the context of each horn of the trilemma for theism in turn.

TM AND THE ATHEISTIC ARGUMENT FROM NO UNSURPASSABLE WORLDS

Here is one reason for thinking that Alternative Worthy History is true: it seems that for some worthy universe, U, God could have altered its history in a way that adds value (for example, by adding more embodied moral agents) so as to produce an alternative worthy universe, U'. If God could have influenced the *history* of some universe to include more value than it did, then that possible but nonactual history is an *alternate* possible universe, as per One History Per Universe. It seems that this alternate universe would be worth creating and sustaining because it differs from U axiologically only by containing more of what is valuable. If at least one worthy universe could have included more of some value, TM cannot include all worthy possible universes; it cannot include both worthy U and its alternative value-added history U'.

Kraay's response to the Atheistic Argument from No Unsurpassable World is to deny NU2 on the basis of his argument that no world distinct from TM can surpass TM. If TM were possible, it would indeed be unsurpassable, as Kraay argues. However, this response to NU2 does not take into account that NU1 is, in fact, a reason to think TM is not possible. One might think that NU1 is true because (a) every possible universe could be improved by the addition of some valuable entity or state of affairs by an omnipotent being; or (b) an omnipotent being could always create additional universes to add value to creation; or (c) every multiverse above some axiological threshold includes at least one universe that an omnipotent being could add value to such that doing so would increase the value of the multiverse.[5] (a) and (c) imply that there is at least one possible universe with two possible alternative worthy histories. By One History Per Universe, only one can be created and sustained by God. Therefore, if (a) or (c) are true, God cannot create and sustain *every* possible worthy universe and TM is impossible. TM is also impossible on (b). (b) may be believed on the basis that a concrete infinity is impossible or on the basis that even if there is an infinite number (of any cardinality) of worthy universes, there is no set of all possible worthy universes because an omnipotent being can always create an additional worthy universe and thereby add value to creation. Kraay's appeal to TM as a possible world does not undermine these ways of supporting NU2; it only assumes they are false. So TM does not refute the Atheistic Argument from No Unsurpassable Universe, contrary to what Kraay claims.

Kraay might respond by insisting that TM already includes every possible addition of value to every universe God would create. Taken literally, this means U' is, after all, included in TM, which is to deny One History Per Universe. This is implausible, as I have argued. Although a universe qualitatively identical to U' may be included in TM, barring bi-location, U' itself cannot be included in TM. The individuals in U who *could have* belonged to a different history of U cannot also actually belong to a universe history different from U.[6]

Alternatively, Kraay might say that it is not axiologically relevant whether U' exists rather than a universe qualitatively identical to U' and that it is better for U and a universe qualitatively identical to U' to exist (along with vastly many other universes qualitatively identical to all the different ways value may be added to U) than for only U (or only U and infinitely many universes qualitatively identical to U) to exist.[7] This reply, however, has several weaknesses. It would require a reformulation of TM (perhaps as a multiverse of all and only universes qualitatively identical to all and only worthy possible universes) that has not yet been examined. It would concede the key premise of the Atheistic Argument from Multiple Surpassable Worlds, MU2: "There are at least two unsurpassable worlds, if there are any". This is because a world with U and a universe qualitatively identical to U' and a world with U' and a universe qualitatively identical to U may both be unsurpassable. It also involves a questionable commitment to near-identical agent world histories.

Let us pause to consider this last important difficulty. It is not implausible on theism that every moral agent is a unique immaterial self and that what it is like to experience life is unique to each person. If this is the case, then even if agents had identical physical bodies, their choices would be experientially different. It is not obvious that choices that are different experientially (or spiritually) can issue in identical behaviors for significant stretches of time, such that the interwoven life histories of billions of people on earth could be identical, or even nearly identical, with those of billions of other agents.

Consider, for example, the multiverse of an infinite space in which all possible configurations of matter are realized.[8] One might suppose that if all possible arrangements of matter are realized, all imaginable qualitatively different histories are realized. However, if arrangements of matter are the result of causal powers, including those of free agents, then which physical arrangements are possible depends upon which choices are possible, and perhaps not all imaginable choices are possible. One reason to think this is that unlike purely material objects, immaterial agents may be thought to have unique essences. We therefore cannot reason, as we may in the case of purely material processes, that for every world history of moral agents there is another set of moral agents who would make identical or nearly identical choices resulting in an identical or nearly identical history, on the basis that there are infinitely many moral agents.

Added to this is the consideration that the precise timing of innumerable events, including the choices of vast numbers of people, is necessary for many significant events, including the precise timing required for the conception of someone as a human person. Minute differences in planet histories can result in very different outcomes, such as the existence of qualitatively physically different humans. Granted, unimaginably small improbabilities of near-identical histories are not impossibilities, and only impossibilities are relevant, but my point is that even if some initial parallels in history may be possible, the near-identical individual life histories required for near-identical universe histories may well be impossible, at least on the theism-friendly assumption that mental life does not supervene on brain activity.

Furthermore, on theism, choices are often responses (or failures to respond) to God and are influenced by God's responses. So both the choices of different agents and God's response to each must be coordinated, perhaps identical, so as to issue in the right context for the same further choices. Even if it is (in some restricted sense) possible that God responds to all of vastly many different agents in different worlds such that their consequent behaviors are identical or nearly so, as required by closely parallel possible world histories, it is not obvious that this is compossible with God's benevolence toward each individual. If each person is a unique self with unique experiences of life, it seems the best specific divine response for one person may well not be identical with the best for any other moral agent, even one identically embodied. The spiritual dimension of the moral choice context, however understood, may be unrepeatable, and this cannot be discounted when considering whether near-identical life histories are possible, given theism. So it may be impossible for there to be universes corresponding to every possible improvement (or axiologically irrelevant alteration) of each universe containing moral agents, especially given theism.

There is a third way that Kraay might respond to my criticism. He might say that if a universe could have been improved by adding value to it, then it cannot be worthy of divine creation in the first place. On this view, a universe that could have value added to it is not worthy of creation because it is never worth creating and sustaining one history *over a better* history. To clarify, this view does not state that a universe must be unsurpassable to be creationworthy; only that it must be unimprovable. These are different because a universe may be unimprovable in the sense that God could not increase the value of the history that began from the initial state of the universe, even if the universe is surpassable in the sense that God could have created and sustained a better initial universe state and history.

This more restrictive notion of creationworthiness, however, again simply assumes that TM is possible without addressing the support for NU2 that challenges this. For example, if every universe could have value added to it by an omnipotent being—alternative (a)—and it is never worth creating and sustaining one universe history *over a better* history, then no universe can be worthy of creation. Similarly, this notion is undermined by the intuition supporting NU2—expressed by alternative (c)—that every multiverse above some worthiness-conferring axiological threshold includes at least one universe to which an omnipotent being could add value. Perhaps every worthy universe is such that God could not have both unsurpassably governed it and added more valuable entities or states of affairs to it, but this claim is an addition to TM. Also, this response does not address the claim—alternative (b)—that God could always add a worthy universe to a multiverse. This response assumes that there is a set of all possible worthy universes, but does not give a reason for thinking so.

TM AND THE ATHEISTIC ARGUMENT FROM MULTIPLE UNSURPASSABLE WORLDS

A second reason for thinking that Alternative Worthy History is true is that it seems God could have caused at least one worthy universe to be slightly different without diminishing or increasing its axiological status. If one worthy universe U could have been different in some trivial way that does not affect its axiological status, such as some difference in an isolated subatomic particle, then that alternative possible universe history, U′ must also be worth actualizing. If so, then TM is impossible: no world can include both U and U′, by One History Per Universe.

Kraay argues against MU2 on the basis of his argument that TM is the unique unsurpassable world, but MU1 actually undercuts Kraay's argument. Again, even if God can create every possible worthy initial universe state, He cannot sustain every possible worthy universe history, given MU1.[9] God cannot, in sustaining the universes He does, actualize both U and U′. So MU1, if true, conflicts with Kraay's PP3, and Kraay has not given any reason to doubt MU1. His argument for TM must say that it is false, but offers no reason to think it is. Moreover, MU1 is *prima facie* true. *If* TM is possible, no world distinct from TM can be unsurpassable, but what MU1 shows is that TM is not possible: no world can include all possible worthy universes (by One History Per Universe). Even if the actual world is an unsurpassable multiverse, it is not the only possible one, by MU1.[10]

This reason for accepting the Alternative Worthy History Thesis avoids the objection, discussed earlier, to the effect that if a universe could have been improved, then it is not worth creating and sustaining. What about the other objection discussed earlier, namely that even though TM cannot include both U and trivial variant U′, this does not matter because TM includes both U and a universe qualitatively identical to U′? The claim that TM includes a universe corresponding to every trivial variation of a universe history entails there are numerous near-identical histories of moral agents. I have argued that this important assumption is questionable. Also, this response concedes that no multiverse can comprise all possible worthy universes and so concedes that TM requires reformulation. Most crucially, if it doesn't matter that a multiverse includes U divinely governed way x and a universe qualitatively identical to U' divinely governed in way y (which is axiologically equivalent to x) rather than including both U and U', then it doesn't matter that, rather than the former, a multiverse includes U' governed way y and a universe qualitatively identical to U governed way x. The fact that these are different and axiologically equivalent means that there cannot be a unique unsurpassable world.

TM AND THE ATHEISTIC ARGUMENT FROM THE SURPASSABILITY OF THE ACTUAL WORLD

The Atheistic Argument from the Surpassability of the Actual World involves the *a posteriori* claim that the actual world contains, or at least probably contains, evil or an absence of good that is inconsistent with God's existence. Though I agree with Kraay that TM does not itself provide the resources to refute this argument, I think that TM achieves less than he thinks it does with respect to apparent absence of possible good (improvability in general) and more than he thinks it does with respect to the problem of evil.

Kraay thinks that for someone to defend this argument by showing that the actual world is surpassable, she must argue either that the actual world fails to contain *all* the universes worth creating and sustaining, or that it fails to contain *only* those universes worth creating and sustaining (or both). Kraay thinks that the former strategy is unlikely to succeed, since it requires surveying and evaluating all actual universes—which are spatiotemporally isolated—in order to determine whether any worthy ones have been left out. But Kraay fails to realize that that there is another way to show that it is false that all worthy universes are included—a way that does not require surveying all universes. One can simply argue that *our* universe could have been governed differently by God while still being worthy of inclusion in TM, or that, even though our universe is unworthy of inclusion, it would have been worthy if governed differently. Given One History Per Universe, this differently governed variant of our universe cannot be included in TM, in which case TM fails to include all worthy universes.

Let's say that a universe exhibits 'governance surpassability' when it contains at least one moment in its history at which God—not agents or nature—could have brought about a better alternative history. In contrast, a universe exhibits 'design surpassability' when God could have improved it in more structurally basic ways, for example by its having a surpassable natural order, or type of moral freedom, or combination of these. Since God can presumably act within a universe's history to alter its fundamental characteristics, this distinction is not sharp, but it is nevertheless important for categorizing *appearances* of surpassability. A governance-surpassable universe's being worthy of actualization may be understood to mean that God would be justified in creating it if, *per impossible,* God need not choose between it and its optimally governed alternative (at each choice juncture). The fact that a universe could have had a better history had it been governed more optimally by an omnipotent being does not mean it is not worthy of actualization in this sense, so our universe may appear governance-surpassable without appearing unworthy of actualization.

To illustrate this distinction, one may think that even if our universe is surpassable because God could have created a natural order or moral agents better than our universe's and even if God may well have created these better universes (so that *this* surpassability is compatible with theism), necessarily God would optimize *our* particular universe with its distinctive (surpassable) potential goods, making its history and narrative the best He can. If it appears He has not done so, this may count against theism. One who thinks God could have done more to make earth and its history better thinks our universe is surpassable even if she grants that its basic design blueprint (its fundamental potencies and parameters, say) may be optimal (or, alternatively, necessary). A universe surpassable in this way may still be creationworthy in the sense of being net good, containing no gratuitous evil, and satisfying other axiological requirements. For example, one may think that all suffering is defeated by good and hence is nongratuitous, but also that God could have prevented the circumstance giving rise to that suffering in a way that would have made our universe *better*. Or one may think that even if all divine permission of evil is justified, God could have guided nature and history so that (say) more favorable natural events, better leaders, and greater happiness emerged. All this may be thought evidence that the actual world is not the greatest possible even if it is a multiverse of worthy universes. Of course, one may include a universe's having no governance surpassability (having no juncture in its history at which God actualized it over a better alternative) as a requirement for being worthy of creation and sustenance, but this does not affect my point that TM shows only that our universe's design surpassability is not evidence against theism, which I think is weaker than the more general conclusion Kraay intends.

One might object that if universes with histories very similar to that of our own universe are possible, including life histories nearly identical to ours (departing from ours at numerous junctures), then the import of this type of surpassability is diminished or even eliminated. The qualitative distinctiveness of our universe, if it has any, would then be more fine-grained and the ways we think it could have been improved may in fact represent distinct universes. If God need not choose between creating a universe nearly identical but better than our universe (or a duplicate of it) and creating our universe (or would choose our universe over that duplicate), and if it is on the whole good that our particular universe exists rather than not, then the fact that God could create a universe nearly identical to ours but with a better history does not by itself count against theism, one may reason. It is better still that God create both and He may well have done so. So, on this near-identical histories multiverse thesis, the surpassability of our universe, by basic design plan or by particular history, is not evidence against theism, one may argue.

In response, recall my earlier argument to the effect that the near-identical histories multiverse thesis is questionable, especially given theism. Also, as noted earlier, even if these nearly identical histories, are possible, TM still requires revision because it cannot include *alternate* worthy histories, by One History Per Universe. Kraay might here reply that only unsurpassably divinely governed universes are genuinely possible because God *necessarily* governs each universe unsurpassably. But if surpassably governed universes are not really possible, our universe's seeming to be governed surpassably may count as evidence against theism, even if it seems overall worth creating.

Are surpassably divinely governed universes possible? This can only be the case if the value added to the multiverse by including these surpassably governed universes outweighs the value of the unsurpassably governed universes they *replace*. Although this is *prima facie* implausible, it may be made intelligible by supposing that God's surpassably governing a universe can be preferable to duplicating (if possible) a universe that is qualitatively identical to its unsurpassably governed alternative. The interuniverse good of diversity or of some other good gained would need to outweigh whatever goods would have been gained by the universe's unsurpassable governance. If such near-identical universes are possible, God's governing some universes surpassably does seem to add many more different types of net good universes to the multiverse, as I will argue in the next section. However, it is not obvious that this is possible. There is the competing consideration that the set of possible optimally governed universes given by God's options for universe design and moral agent creation would not require supplementation by the possibilities afforded by suboptimal governance such that the loss of good the latter entails is outweighed. Thinking otherwise may seem to diminish God's omnipotence. In any case, if near-identical moral agent histories are not possible, as I have argued, then God would not surpassably govern any universe, for that would be to choose a gratuitous loss of good.

TM AND THE PROBLEM OF EVIL

Now let us consider the relationship between TM and the apparent surpassability of our universe due to the evil it contains. Kraay thinks that TM merely reframes, but does not advance, discussion of the problem of evil. He correctly points out that our universe's seeming to exhibit certain bad-making properties, such as evil God would be unjustified in allowing, may be evidence this is not the greatest possible world, even if it is granted that a divinely furnished multiverse can include surpassable universes. Although I agree that the epistemic or logical possibility that we live in a multiverse does not solve the problem of evil, I think it substantially enhances theistic responses. This is because whether or not God's permission of an evil is necessary for a greater good depends upon what the alternatives to that permission are, and these alternatives are different on the single-universe view than on the multiverse view. This is especially the case for the near-identical universes multiverse view.

It is not obvious that one could be warranted in believing that an evil is gratuitous if one thinks a near-identical histories multiverse is epistemically possible, unless it is reasonable to think our universe is not net good.[11] Given a multiverse of near-identical universes, God's permitting an evil in a net good universe may not be necessary for an offsetting good *in* that universe, but it seems that if God did not permit that evil, or one qualitatively identical to it in another universe, the multiverse would contain at least one less universe of net good and therefore would not be the best possible multiverse.

For example, consider net good universe U and net good universe V, which are qualitatively identical until they differ by God's permitting an evil *e* in U that is not necessary for any greater good in U (and not permitting what would have been a qualitatively identical evil in V). If God prevents *e* in U, and in all universes qualitatively identical to U (up to that moment), these universes would have continued to be qualitatively identical to V, at least up to some further juncture. But then there would have been vastly fewer net good different universes in the multiverse. U and all universes qualitatively identical to U (up to and including the moment *e* occurs) would not exist. Since the multiverse could still have contained just as many universes qualitatively identical to V up to that further juncture had God permitted *e* instead, it seems that for God to prevent *e* and its counterpart in all universes is for God to choose a gratuitous loss of good.[12] A multiverse that includes U and all universes qualitatively identical to U, in addition to V universes, seems to be a greater multiverse than one that does not. More simply, the net good of U outweighs the evil *e* and requires God's permission of *e*. Therefore *e* cannot be a gratuitous evil.[13]

Bradley Monton (2010) has argued against this view. He compares net good universe U containing a gratuitous evil *e* to universe Nice-U that is identical to U in all respects except that Nice-U does not contain *e*. According to Monton, the fact that *e* distinguishes U from other universes does not mean God would create it to add value to the multiverse. He thinks God would create and sustain a duplicate of Nice-U rather than create and sustain U.

Even if this scenario is possible and Monton is correct, it remains true that the requirement for apparent gratuitousness is stronger on the multiverse view than on the single-universe view. On the single-universe view, an evil may be considered gratuitous if and only if its permission by God seems unnecessary for a greater good *within the universe*, whereas on the multiverse view, an evil may be considered gratuitous if and only if its permission by God seems unnecessary for a greater good *within the multiverse*. The latter requires its seeming that God's permission of the evil makes no other significant difference to the history of our universe, whether good or evil or neither (or the universe is judged to not be on the whole good). If the evil does make such a difference, which seems often the case, God's permission of it is necessary for the good of there being such a universe in the multiverse.

Even if one thinks that near-identical universe histories are not possible, the requirement for apparent gratuitousness is stronger on the multiverse view than on the single-universe view in other ways. An evil may seem gratuitous because although it seems that God's permission of that *type* of evil is necessary for some good, it also seems that the same amount or more of that good, or at least more net good, could have been achieved with less of that type of evil. For example, one might think that although divine permission of moral evil is necessary for the greater good of our having significant moral freedom, God would not have permitted the worst of moral evils of our world because it would have been better for our moral freedom to have been curtailed to a greater degree. Likewise, one might think that although God must permit natural evil to preserve the uniformity of the natural order, God would have sacrificed more of the uniformity of nature in order to prevent more natural evil. These judgments are compatible with there being no gratuitous evil on the multiverse view but not on the single-universe view. A proponent of TM may hold that it is good for there to exist a variety of goods, and that as a consequence, to maximize the value of the multiverse, God must actualize a variety of design plans or general policies with respect to goods and their costs. She may grant that an optimal trade-off between the existence of significant morality and divine prevention of moral evil would involve less freedom and less moral evil than we find on earth, but note that it may be good for the multiverse to contain moral communities with our kind of moral freedom in addition to moral communities with a more restricted moral freedom.[14] On the other hand, if there is only one universe, there is no explanation for why God did not choose the better trade-off between moral freedom and divine permission of moral evil.[15]

This does not mean that, given the multiverse view, one cannot be warranted in judging that less moral or natural evil should have been permitted by God. It just means that one must consider the multiverse-wide perspective in order to make this judgment. Even to judge that a trade-off between global goods and evils on earth could have been better is to make a judgment on a vast scale. It is an estimation of the value of very general goods such as humans having significant morality and nature's being highly regular and the comparative disvalue of the totality of broad categories of evils they require, such as innocent suffering. These judgments cannot be made very finely. The perceived difference between a better possible relation between these goods and their costs and the actual relation must be correspondingly vague and large-scale. For God to adopt a correspondingly significantly different policy with respect to these global goods would be for God to create a very different universe than ours.

An evil may also seem gratuitous because although it seems that God's permission of that type of evil is necessary for some good, it also seems that the same amount or more good could have been achieved with less of that type of evil because *different* goods than the suggested theodical goods, involving better or no trade-offs, could have been achieved instead. For example, one might think there is gratuitous moral evil because God didn't have to create free agents like us who often make moral mistakes and need to learn virtue by experiencing evil; he could have created better agents. Likewise, one might think that natural evil is gratuitous because an omnipotent being could have created a universe with moral agents and a different set of natural laws that would not have produced natural evils. This also is compatible with there being no gratuitous evil on the multiverse view but not on the single-universe view. On the single-universe view, this is evidence for gratuitous evil: why didn't God pursue the better different type of trade-off instead of permitting these evils? On TM, it may be that different goods *have* been pursued in addition to the goods of the considered trade-off. A proponent of the TM can respond that God may have created and sustained universes with moral agents who do not require the lessons given by suffering, and may have created universes without natural evil, but that it is optimal for God to also create the distinct good of a universe like ours.[16]

CONCLUSION

I have challenged Kraay's TM by pointing out that if God has a choice between alternative worthy universe histories, TM cannot include *all* possible worthy creatable universe histories. Kraay may respond that only unsurpassably governed universe histories are worthy of creation and sustenance.[17] If this is true, there may be only one worthy universe history for each divine governance choice. Even so, TM still faces the challenge posed by atheistic arguments from No Unsurpassable World that an unsurpassably governed multiverse is impossible because it can always be improved, and the challenge posed by atheistic arguments from Multiple Unsurpassable Worlds that God must choose between unsurpassable multiverses (if there are any) because any multiverse could have been different in an axiologically irrelevant way. So TM does not refute these arguments, contrary to what Kraay claims. I have also argued that Kraay is mistaken in thinking that because TM includes surpassable universes, the apparent surpassability of our universe is not evidence against theism as long as our universe is creationworthy. Although our universe's apparent design surpassability would not be evidence against theism, any apparent divine governance surpassability still would be.

In the last section I argued, against Kraay, that the multiverse view does impact the problem of evil. Whether God's permission of an evil is necessary for a greater good or not depends on what the alternatives to that permission are. If creation includes near-identical universe histories, God's permission of an evil may be necessary for the existence of a worthy universe though unnecessary for a greater good within the universe containing the evil. Such an evil would be gratuitous if the universe containing the evil were the only universe—creation would have been better without it—but not gratuitous on the multiverse view. Apart from near-identical universe histories, which I argue against in the case of moral agent histories, the multiverse view still narrows the class of evils that can be reasonably judged gratuitous. An optimal multiverse may contain a variety of trade-offs between good and evil, some of which would be suboptimal if there were no other better trade-offs in creation. For example, it may be optimal for creation to contain universes with a natural order that gives rise to pain and suffering, as ours does, as well as universes with a natural order that does not. In this case the existence of natural evil is not gratuitous (though its amount may be). On the single-universe view, if universes with a natural order that gives rise to moral agents but no natural evil are thought to be possible, the existence of natural evil may indeed be gratuitous.[18]

NOTES

1. Kraay does not commit to the inclusion or exclusion of entities that are neither worthy nor unworthy of actualization, if there are any.
2. NU1 is not equivalent to NU2 because God may actualize a better world by either *improving* a universe or multiverse, or by creating a better type of universe or multiverse initial state than He has.
3. This premise might be challenged on the basis that certain simple possible worlds, including the world in which God does not create anything, could not be different without affecting its axiological status. MU1's scope could easily be restricted to avoid this objection.
4. Either Kraay objects to Alternative Worthy History and says that U' is not possible or he objects to One History Per Universe and says that both U and U' are included in TM. Either way he must claim either that God could not have governed differently than He has or that all alternatives to God's actual governance include unworthy universes. That disjunction must be considered part of TM.
5. It is important to note that Kraay is neutral concerning the truth of NU1 as well as these possible replacements. See Section 4 of Kraay (2010).
6. As Kraay notes (2010, 363 n34), Peter van Inwagen and Tom Talbott have objected in conversation that TM cannot include all worthy universes because individuals exist in more than one worthy universe but cannot bilocate. Kraay there replies by pointing to his independently motivated "modal collapse" argument to the effect that if theism is true, TM is the only possible world there is (Kraay 2011). But Kraay's modal collapse defense does not undermine NU1 or its implication that TM is not possible. On universe-bound individuals, see also Lougheed (forthcoming).
7. Qualitatively identical universes are not duplicate universes. See Bradley Monton (2010) on the latter.
8. Matter is thought to have the same essential properties in every similar universe, so if the possible states of matter do not vary continuously they may be considered finite in number. All possible states of such matter may therefore be thought actual and duplicated in an infinity of universes.

9. MU1 should be understood as "For any possible way creation could have been, *that creation* could have been different in an axiologically irrelevant way" and not simply as "There are different possible axiologically equivalent ways creation could have been". There being different possible axiologically equivalent initial universe states is sufficient for the latter. These need not be thought of as mutually exclusive alternatives, but different axiologically equivalent ways God could have sustained and governed a universe or multiverse must be. Also, MU1 has the implicit qualification that *God* can bring about this axiologically equivalent way creation could have been (instead of this being brought about by creaturely freedom or indeterministic processes).
10. My own view is that MU1 is unwarranted. TM does not suggest this, however.
11. I defend the standard view that God and gratuitous evil are incompatible in Schrynemakers (2013).
12. God's sustaining U by permitting e does not mean God could not create as many universes qualitatively identical to V as He would have if He had prevented e.
13. U would have had a better history if God had prevented e, so U is surpassably governed by God. As my argument here implies, unsurpassable governance of the multiverse plausibly requires surpassable governance of universes on the near-identical histories multiverse view. If near-identical moral agent histories are impossible, then divine surpassable governance of universes is not required.
14. God's governing a universe by a general policy, or global, trade-off between good and evil that would have been sub-optimal if that were the only universe does not mean that universe is surpassably governed on the multiverse view. I take divine trade-offs between broad types of goods and evils to be part of the design plan of a universe, determining its potential goods.
15. The single universe view can incorporate some of these considerations by positing other inhabited planets in our universe.
16. For a similar argument in this vein, see Hudson (2013).
17. Unsurpassable governance may be relative to the multiverse rather than to each universe.
18. I thank Klaas Kraay for very helpful comments on earlier drafts of this chapter.

REFERENCES

Hudson, H. 2013. "Best Possible World Theodicy." In *The Blackwell Companion to the Problem of Evil,* edited by D. Howard-Snyder and J. McBrayer, 236–50. Malden, MA: Wiley-Blackwell.
Kraay, K. 2010. "Theism, Possible Worlds, and the Multiverse." *Philosophical Studies* 147: 355–68.
Kraay, K. 2011. "Theism and Modal Collapse." *American Philosophical Quarterly* 48: 361–72.
Kraay, K. 2012. "The Theistic Multiverse: Problems and Prospects." In *Scientific Approaches to the Philosophy of Religion,* edited by Y. Nagasawa, 142–162. Houndsmills, UK: Palgrave MacMillan.
Kraay, K. 2013. "Megill's Multiverse Meta-Argument." *International Journal for Philosophy of Religion* 73: 235–41.
Lougheed, K. Forthcoming. "Divine Creation, Modal Collapse, and the Theistic Multiverse." *Sophia.*

Megill, J. 2011. "Evil and the Many Universes Response." *International Journal for Philosophy of Religion* 70: 127–38.

Monton, B. 2010. "Against Multiverse Theodicies." *Philo* 13: 113–35.

Schrynemakers, M. 2013. "God and Gratuitous Evil: General Policy Theodicies and the Illusion of Gratuitous Evils." PhD dissertation, CUNY Graduate Center.

Turner, D. 2003. "The Many-Universes Solution to the Problem of Evil." In *The Existence of God*, edited by R. Gale and A. Pruss, 1–16. Aldershot, UK: Ashgate.

7   Best Worlds and Multiverses

Michael Almeida

Multiverse cosmologies promise solutions to some intractable problems in philosophical theology, in particular the problem of evil and the problem of suboptimality.[1] Assuming the actual world is a multiverse, theists can concede that our universe U is a suboptimal universe—as it evidently is—but affirm that U is necessary to the actualization of the best possible world. Theists can also concede that there are evils E that serve no greater purpose in U—as there evidently are—but affirm that E is necessary to the actualization of the best possible world.

Since every instance of evil in every universe is necessary to the actualization of the best possible world, there is no pointless evil in any universe. There is certainly evil that is unnecessary to any greater good *in the universe,* and this gives the illusion that there is pointless evil. But the problem of evil is resolved.[2]

Further, there are no better possible worlds than the actual world. Our world is not suboptimal. There are certainly universes that are better than ours, and this gives the illusion that our world is suboptimal. But the problem of suboptimality is resolved.

Multiverse solutions to the problem of evil and the problem of suboptimality depend on modal theses M1 and M2:

M1. Necessarily, there is a world that includes every universe whose overall value is n or greater, for positive n.[3]

M2. Necessarily, God actualizes a world that includes every universe whose overall value is n or greater, for positive n.

The metaphysical thesis in M1 entails that every possible universe whose overall value is n or greater (for positive n) is included in some single world, W. If possible worlds might overlap with respect to universes, then some universes U in W might also be included in other worlds W′. If possible worlds do not overlap with respect to universes, M1 entails that there is a world that includes a duplicate of every universe whose overall value is n or greater.[4]

It is difficult to see how multiverse solutions to the problem of evil and the problem of suboptimality might depend instead on the weaker modal theses. Consider, for instance, the thesis in M1′:

M1′. Necessarily, there is a world that includes every universe whose overall value is *n* or greater, for *some* positive *n* or other.

The thesis in M1′ entails that every possible universe whose overall value is *n* or greater for some positive *n* (or other) is included in some single world, W′. But we cannot be sure that W′ is a best possible world without assuming an arbitrarily restricted principle of recombination.[5] For instance, let W′ include duplicates of every universe whose overall value exceeds some large finite value. W′ is a best possible world only if the principle of recombination rules out worlds that include duplicates of every universe whose overall value exceeds some slightly less large finite value. But there is no nonarbitrary reason to believe that the principle of recombination is suitably restricted.

A world that includes a duplicate of every universe whose overall value is on-balance good is a best possible multiverse.[6] Let's say that the set of universes that are included in the best possible world is the set of good enough universes. M2 entails that, necessarily, God actualizes the world that consists in all of the good enough universes. If M1 and M2 are true, then the problem of evil and the problem of suboptimality are easily resolved.

My aim is to show that thesis expressed by M2 is false. It is possible that God does not actualize the best possible world. Indeed, it is true in every world that God might have actualized a suboptimal world. The argument I advance does not assume that God lacks any divine attribute in any world in which he exists. God's attributes include essential omnipotence, essential omniscience, essential perfect goodness and necessary existence. So, God has those properties in every possible world. The argument does not assume that there are any possible worlds in which God is unable to actualize the set of good enough universes.[7] It is a necessary truth that God can actualize a world that includes all and only the good enough universes.

Further, the argument makes no assumptions concerning the nature of free will. Libertarian accounts of free will might be correct, but strong or weak compatibilist accounts of free will might also be correct. The argument makes no assumptions concerning the principle of alternative possibilities.[8] It is consistent with the argument that agents are morally responsible for what they do, even when they are unable to do otherwise. Morally significant actions have moral value even when every moral agent is unable to go wrong.

If M2 is false, then, necessarily, God might not actualize a world that includes every positive-valued universe. Indeed, I aim to show that God might not actualize a world that includes every positive-valued universe though God *can* actualize a world that includes every positive-valued universe. I conclude that multiverse solutions to the problem of evil and the problem of suboptimality fail. But that's only because we do not need a solution to the problem of evil and the problem of suboptimality.[9]

SOME PRELIMINARIES

A possible world is a maximal or complete state of affairs. A state of affairs W is maximal just in case it is true, for each state of affairs S, that W includes S or W excludes S. Let's call the (full or partial) instantiation of a possible world W a universe or cosmos, U of W. Let's assume that a world W might include several states of affairs U each of which constitutes a distinct universe of W. We allow that the universes in W might be 'island universes' causally isolated from all other universes in W.[10] But, as we understand 'possible worlds', the (indexical) phrase 'the actual world' uttered in any island universe U of W refers to W, and not merely to the isolated part of W in which the phrase is uttered.[11]

If a world W includes more than one universe, the universes together constitute the cosmoi or multiverse M of W. Since a universe of W might be a partial instantiation of W, U might not be an exhaustive instantiation of the states of affairs in W.[12] We assume that no world includes more than one multiverse. So, multiverses M of W must be full or complete instantiations of W.

We'll say that God actualizes a world W just in case either God causes every contingent state of affairs in W or God brings about the largest state of affairs that he can unrestrictedly or strongly actualize in W.[13] A world obtains just in case its universe(s) exist(s).[14]

Let's stipulate that there is a best possible world, W_B, that includes all of the universes, or duplicates of all of the universes, whose overall value is positive. We make no assumptions about the number of universes in W_B. W_B might include a finite number of universes, where each U in W_B has an overall value of n or greater for positive n. W_B might include infinitely many universes each of which has a value of n or greater. W_B might include uncountably many improving universes, each of which is n or greater in value.

If the best possible world includes every universe whose overall value is n or greater, then it includes some universes that are morally perfect and some universes that are naturally perfect.[15] Some regard the natural perfection of a world as easier for God to achieve, since God can strongly actualize a naturally perfect world. In morally perfect universes, moral agents always freely go right with respect to their morally significant actions.[16] The morally good states of affairs in universes are the result of moral agents freely going right with respect to morally significant actions and the morally evil states of affairs in universes are the result of moral agents freely going wrong with respect to morally significant actions.

Moral agents in perfect universes are free in a sense relevant to being morally responsible for what they do. We assume that morally significant actions have moral value only if moral agents are free in the sense relevant to being morally responsible for what they do. We make no assumptions about the proper account of free will. If libertarian accounts of free will are correct, then moral agents in perfect universes are libertarian free. If a strong or weak compatibilist account of free will is correct, then moral agents in perfect universes are weak or strong compatibilist free.

We grant that there are no worlds in which God lacks the power, knowledge, or goodness necessary to actualize the best possible world. There are no worlds at which the set of actualizable worlds is distinct from the set of possible worlds. God can unrestrictedly actualize the best possible world.[17]

THE ARGUMENT AGAINST M2

We have stipulated that there is a best possible world, W_B, that includes all of the universes, or duplicates of all of the universes, whose overall value is on-balance positive.[18] This is the assumption in M1 and is regarded as important to the multiverse solutions to the problem of evil and the problem of suboptimality.

1. W_B is the best possible world.

Since we have assumed that there are no limitations on God's power to actualize worlds, it is necessary that God can actualize W_B.

2. Necessarily, God can actualize W_B.

Let's suppose that God's attributes are inconsistent with his failing to avail himself of the option of actualizing W_B.[19] If God's attributes are consistent with his failing to avail himself of the option to actualize W_B, then there is nothing especially problematic in God's actualizing a suboptimal world. Indeed, there is nothing especially problematic in God actualizing worlds that include unjustified evils.[20] God does not actualize any world that is less than the best unless he lacks either the power or goodness or knowledge to actualize the best world.

3. Necessarily, God can actualize W_B only if God does actualize W_B.

It follows from (2) and (3) that God must actualize the best possible world W_B.

4. Necessarily God actualizes W_B.

Since W_B is the best possible world, it will include all universes that are on-balance good, among them many universes that are morally perfect.

Morally perfect universes contain no moral evil, but they vary in the amount of moral good they include. A morally perfect universe might include only a few moral agents, each of whom performs only a few small acts of beneficence. The universe would be morally perfect but it would not contain much moral good. Another morally perfect universe might include many moral agents who are always going right with respect to many large acts of beneficence. The universe would also be morally perfect, but it would of course contain much more moral value.

But most of the moral value of morally perfect universes is due to moral agents observing moral prohibitions against the violation of individual rights. These are moral agents fulfilling the typically negative duties that form the fundamental requirements of justice. It's widely agreed, even among consequentialists, that the demands of justice are the weightiest or most important requirements of morality. Moral theorists as diverse as Hume (1981), Kant (1953), Rawls (1975), Nozick (1975), Gauthier (1987), and Cohen (2008) all agree on the relative importance of the requirements of justice. The requirements of justice prohibit the violation of basic moral rights such as the right to life, the right to freedom, and perhaps political, property and economic rights.

Call the set of morally perfect universes included in the best world U_B. U_B includes those universes where every moral agent always goes right with respect to the requirements of beneficence and justice. The agents in those universes not only maximize beneficence but they also constrain their behavior in ways that ensure the observance of the rights of all. Since the requirements of justice are the most important moral requirements, the universes in U_B are extremely valuable. Indeed, we know that W_B includes every extremely valuable, morally perfect universe in U_B.

Conversely, we know that W_B includes none of universes in which every moral agent always goes terribly wrong with respect to the requirements of beneficence and justice. Call that set of universes U_E. The agents in the universes in U_E not only maximize maleficence but also violate all of requirements of justice.[21] Since the requirements of justice are the most important moral requirements, the universes in U_E are extremely disvaluable.

5. W_B includes every extremely valuable, morally perfect universe $U_B = \{U_{B0}, U_{B1}, U_{B2}, \ldots, U_{Bn}\}$ in which all moral agents always observe the requirements of justice and beneficence and W_B excludes every extremely disvaluable, morally imperfect universe $U_E = \{U_{E0}, U_{E1}, U_{E2}, \ldots, U_{En}\}$ in which all moral agents always violate the requirements of justice and beneficence.

According to premise (4), God necessarily actualizes W_B. It follows from God's nature, presumably, that he cannot fail to avail himself of the option of actualizing the best world W_B. And it follows from unrestricted actualization that there is no world in which God does not have the option of actualizing

W_B. So, the set of universes in W_B not only exhaust the actual universes that obtain, they exhaust the *possible* universes that exist. Premise (4) entails that the set of actual universes just is the set of all possible universes.

According to premise (5), W_B includes all of the extremely valuable, morally perfect universes in U_B and W_B contains none of the extremely disvaluable universes in U_E. And according to premise (4), W_B includes all possible universes. (4) and (5) are both true only if W_B includes all of the *most valuable* universes, none of the *least valuable* universes and W_B includes all of the *possible* universes.

But (4) and (5) are not consistent. It can be shown that if W_B includes all of the most valuable universes in U_B and none of the least valuable universes in U_E, then, if W_B is the best possible world, then premise (4) is false. W_B does not contain all of the possible universes. Indeed, W_B is the best possible world and includes all and only universes in U_B only if W_B excludes some genuinely possible, *extremely bad* universes.[22]

This presents a serious problem for multiverse solutions to the problem of evil and the problem of suboptimality, since it entails that M2 is false. It is simply not necessary that God actualize the best possible world W_B. It is possible that God can actualize W_B, but actualizes instead an extremely bad, suboptimal world.

Propositions (4) and (5) Are Inconsistent

As we have noted, much of the moral value of the universes in U_B is the result of moral agents observing requirements to constrain their behavior within the bounds of justice. So, much of the moral value of the universes in U_B is the result of moral agents freely choosing not to actualize worlds that include morally unjust universes.

Compare the universe U_{E0} in U_E and the universe U_{B0} in U_B. Let U_{E0} be a universe in which all of the moral agents in U_{B0} go terribly wrong with respect to the requirements of justice and beneficence. We know that there is such a universe, since much of the value in U_{B0} is the result of moral agents freely deciding not to go terribly wrong.[23] We may let U_{E0} be a universe of vast and profound injustice, a universe of horrendous suffering and extreme harm. It is among the most stringent requirements of justice in U_{B0} that moral agents are prohibited from bringing about any world that includes the universe U_{E0}. And by hypothesis the moral agents in the morally perfect universe U_{B0} are strictly conforming to those stringent requirements. Universal conformity to the requirements against actualizing any world that includes U_{E0} fulfills some of the most important requirements of justice in U_{B0} and so contributes a great deal to the moral value of U_{B0}.

6. The fact that moral agents in U_{B0} all conform to the requirements not to actualize any world that includes U_{E0} contributes to the moral value of U_{B0}.

But universal conformity to moral requirements not to actualize any world that includes U_{E0} contributes to the moral value of U_{B0} only if it is at least *metaphysically possible* for moral agents in U_{B0} to actualize a world that includes U_{E0}.[24] If the moral agents in U_{B0} conform to requirements not to actualize any world including U_{E0} as a matter of metaphysical necessity, then it is uncontroversially true that their conformity to these requirements is not freely exercised. And if universal conformity to the requirements of justice is not freely exercised, then such conformity does not contribute to the moral value of U_{B0}.

7. Conformity to requirements against the actualization of a world that includes U_{E0} contributes to the moral value of U_{B0}, only if it is metaphysically possible for moral agents in U_{B0} to actualize a world that includes U_{E0}.

Suppose it is metaphysically impossible for moral agents in U_{B0} to actualize a world that includes the universe U_{E0}. In that case it is of course metaphysically necessary that all moral agents in U_{B0} conform to the prohibitions against the actualization of a world including U_{E0}. No matter what the moral agents in U_{B0} might do, those agents necessarily conform to the requirements against actualizing a world that includes U_{E0}. But if it is metaphysically necessary that all of the agents in U_{B0} conform to the requirements against the actualization of U_{E0}, then, on any account of the nature of free will, no moral agent in U_{B0} freely conforms to the prohibitions against the actualization of U_{E0}.

Consider a strong compatibilist account of free will. According to strong compatibilists, alternative possibilities (of action) are necessary for free will and moral responsibility.[25] So, strong compatibilists are committed to a version of the principle of significant freedom (PSF).[26]

PSF. A moral agent S is significantly free with respect to A only if S could have done other than A.

A person is significantly free relative to action A only if S is free in the sense relevant to S's being morally responsible for A. PSF entails the weaker principle of genuine possibilities (PGP). No moral agent is significantly free with respect to any action unless she satisfies PGP.

PGP. A moral agent S is significantly free with respect to A only if there is a metaphysically possible world in which S does other than A.[27]

According to strong compatibilism, if a world including U_{E0} is not metaphysically possible, then the moral agents in U_{B0} do not freely comply with the moral restrictions not to actualize a world including U_{E0}. But then universal conformity to the restriction against actualizing any world that includes U_{E0} does not contribute to the moral value of U_{B0}.

Consider instead weak compatibilism. According to weak compatibilists, moral agents are free in the sense relevant to moral responsibility only if those agents exhibit the proper sort of control with respect to their actions. But a moral agent has the proper sort of control with respect to an action A only if there is a possible world in which the agent is has sufficient reason to do other than A and he does otherwise.[28] So, no moral agent has the proper sort of control unless he satisfies PGP. If moral agents in U_{B0} do not satisfy PGP, then universal conformity to the restriction against actualizing any world that includes U_{E0} does not contribute to the moral value of U_{B0}.

Finally, of course, on libertarian accounts of free will, a moral agent S is relevantly free with respect to action A only if S is able or has it within his power to do other than A. So libertarian accounts of free will are also committed to both PSF and PGP.

On any account of free will, the moral agents in U_{B0} freely conform to the requirements against the actualization of a world including U_{E0} only if it is at least metaphysically possible that the moral agents in U_{B0} actualize a world including U_{E0}. Conformity to moral requirements against the actualization of a world including U_{E0} contributes to the moral value of U_{B0}, only if it is metaphysically possible for moral agents in U_{B0} to actualize a world including U_{E0}.

In the morally perfect universe U_{B0}, the fact that moral agents conform to requirements against the actualization of a world including U_{E0} constitutes a very important source of moral value. Indeed, adherence to the requirements of justice is the most important source of moral value in U_{B0} as it is in any universe. So, it must be at least metaphysically possible for moral agents in U_{B0} to actualize U_{E0}.

8. It is metaphysically possible that moral agents in U_{B0} actualize a world including the extremely unjust universe U_{E0}.[29]

But it's metaphysically possible that moral agents in U_{B0} actualize a world including the extremely unjust universe U_{E0} only if U_{E0} is included in W_B. Recall that, according to (4), the set of actual universes in W_B just is the set of all possible universes.

9. The universe U_{E0} is included in W_B.

But recall also that, according to premise (5), W_B includes none of the extremely disvaluable universes in U_E. So W_B does not include U_{E0}. U_{E0} is a universe of vast and profound injustice, a universe of horrendous suffering and harm. It is among the most stringent requirements of justice in U_{B0} that moral agents not actualize U_{E0}.

10. The universe U_{E0} is not included in W_B.

(9) and (10) constitute an explicit contradiction. They cannot both be true. We know by premise (5) that W_B includes all of the universes in U_B and none of the universes in U_E. But we found that U_{B0} is included in W_B only if U_{E0} is included in some metaphysically possible world. The moral value of U_{B0}

is in part the result of moral agents conforming to the requirement not to actualize those metaphysically possible worlds that include U_{E0}.

Therefore, there must be a possible world W' diverse from W_B that includes U_E. But if there is a metaphysically possible world W', then premise (4) is false and so is the thesis in M2. God does not necessarily actualize W_B. But there are additional consequences following from the rejection of (4). If (4) is false, then so is premise (3). We have granted that God can unrestrictedly actualize the best possible world W_B, and so it is true in every world that God can actualize W_B. But we must reject (3). It is not a necessary truth that God can actualize the best possible world only if God does actualize the best possible world.

CONCLUDING REMARKS

According to multiverse solutions to the problem of evil and the problem of suboptimality, there is a world W_B that includes all of the on-balance good universes—or duplicates of all of the on-balance good universes—and God necessarily actualizes that world.

We noted that multiverse solutions to the problem of evil and the problem of suboptimality depend on theses M1 and M2. M1 ensures that there is a best possible world and M2 ensures that God actualizes *only* the best possible world.

M1. Necessarily, there is a world that includes every universe whose overall value is n or greater, for positive n.

M2. Necessarily, God actualizes a world that includes every universe whose overall value is n or greater, for positive n.

We noted that if M1 and M2 are true, then the problem of evil and the problem of suboptimality are easily resolved. Every instance of evil in every universe in W_B is necessary to the actualization of the best possible world and the greatest amount of goodness, so there is no pointless evil in any universe. Further there are no better possible worlds than W_B, since there are no other worlds than W_B. So, our world is optimal among worlds.

But we showed that M2 is false. Indeed it is simply not possible that, in every possible world, God actualizes a world that includes every universe whose overall value is n or greater, for positive n. We grant that there is a world W_B that includes all and only universes whose value is n or greater for positive n. W_B will no doubt include, among other universes, all of the best universes in U_B. We argued that, among the universes in U_B, there are morally perfect universes such as U_{B0} whose moral value is due to universal conformity to the requirements of justice and beneficence. We argued that universal conformity to requirements prohibiting the actualization of universes such as U_{E0} does not contribute to the moral value of U_{B0} unless there are worlds that include universes such as U_{E0}. And we concluded that if W_B includes all of the best universes in U_B, and none of the worst universes in

U_E, then there are worlds diverse from W_B that include all or some of the universes in U_E.

If there is a world W_B and diverse worlds W', then God does not necessarily actualize W_B. God possibly Will Defense depends actualizes W' which includes some of the worst possible universes in U_E. The thesis in M2 is false, and indeed it is necessarily false. It is necessarily true that God cannot actualize the best possible world W_B unless it is possible for him to actualize much worse possible worlds W'. Since M2 is necessarily false, multiverse solutions to the problem of evil and the problem of suboptimality fail. The good news is that, since M2 is necessarily false, there is no problem of evil or problem of suboptimality to resolve.[30]

NOTES

1. What I am calling the 'problem of evil' is really a set of problems related in complex ways. The basic problem is that there is pointless and eliminable evil or it seems very probable that there is pointless and eliminable evil. And the existence of such evil is not consistent with the existence of an Anselmian God. What I am calling the 'problem of suboptimality' is sometimes called the 'problem of less than best'. The basic problem is that the actual world is very probably not the best possible world. And the existence of a suboptimal world is not consistent with the existence of an Anselmian God.
2. The multiverse solution to the problem of evil might seem to turn on the question of whether our universe is good enough to be included in a multiverse. This is a red herring for a variety of reasons. A multiverse that solves the problem of evil can have universes of any value whatsoever. What matters is that the evil in those universes is justified and that the multiverse is optimal. Neither of these depends on the overall value of the universes.
3. It is not crucial that n be positive, so long as the multiverse is the exclusively possible world.
4. Multiverse solutions to the problem of evil and the problem of suboptimality are confused in a variety of ways that I don't pursue. It is never observed, for instance, that if God necessarily actualizes a multiverse M then M is the best possible world, *no matter how bad the universes are that compose M.* The multiverse solution to the problem of evil, for instance, would be resolved if M were composed of universes that included only horrific moral and natural evils. Each one of those evils would be necessary to the actualization of the best possible world, M. The value of universes that compose M is irrelevant to the theodicist's goals. It is also never observed that God cannot choose the universes to instantiate in M, since there are no possible universes except those in M, according to multiverse theodicies.
5. Cf. Lewis (1986, Section 1.8).
6. Might there be on balance good worlds that God cannot actualize? Klaas Kraay reminds me that some universes might be on balance good but contain large amounts of terrible evil. Perhaps God could not actualize such worlds, but this would invoke another principle of moral perfection besides the requirements of choosing optimal worlds and worlds without gratuitous evil.
7. Recall that Alvin Plantinga's Free Will Defense depends on its being possible that the set of actualizable worlds is not the set of all possible worlds. I do not make any such modal claim. See Plantinga (1974, esp. Section IX).

8. There is a large body of literature and various more or less useful analyses of the principle. The most important discussions begin with Frankfurt (1969).
9. It is false both that God cannot coexist with gratuitous evil and that God must actualize an optimal world. See Almeida (2012).
10. See Bricker (2001) and Lewis (1986, Section 1.6). Lewis does not allow for a single world to have spatiotemporally isolated parts. For the sake of discussion, I am conceding that there might be such worlds.
11. For persuasive reasons that 'the actual world' is not an indexical, see van Inwagen (1980).
12. The issue is more complicated. It is true for each universe U of world W that the states of affairs in U are maximal. U is such that, for every state of affairs P, U includes or precludes P. In short, for each universe U of W, every proposition p is either true or false in U. We'll say that a proposition p is true in a world W just in case p is true in a universe U of W. Additional assumptions are necessary to ensure consistency in worlds including the assumption that individuals are universe-bound.
13. For the differences among strong actualization, weak actualization and unrestricted actualization, see Almeida (2012, Chapter 4).
14. I take universes to obtain iff. they exist, since they are instantiations (perhaps partial instantiations) of worlds.
15. It should be noted that near-perfection would serve the argument just as well. By 'a naturally perfect world' I mean a world that includes no natural evil.
16. An action A is morally significant for an agent S iff. it would be morally right for S to perform A and wrong for S not to perform A or vice versa. See Plantinga (1974, 166ff.).
17. See Almeida (2012, 108ff.).
18. As noted above in note (2), the foregoing sentence, though typical in the exposition of multiverse solutions to the problems of evil and suboptimality, is strictly speaking not coherent. We will learn that there is only one possible world, so such a world *could not* contain duplicates of any universe in any world except itself.
19. For prominent examples, see Mackie (1955) and Rowe (2002).
20. There is insufficient discussion of this point in the literature. It's often assumed without much argument that God cannot actualize a world that includes a single instance of gratuitous evil but God might actualize a suboptimal world. God might be permitted to actualize a suboptimal world, it is urged, because there might not be a best possible world. But the problem is much more difficult than that. Consider C.

> C. For any world W in which there is no evil, there is a better world W′ in which there is some evil.

If there is no best possible world, then C is likely true. But the problem is even more difficult. The argument that God can actualize a world with no evil W when there is a better world with some evil W′ cannot appeal to *equality principles*, because C′ is also likely true.

> C′. For any world W in which there is no evil, there is a world W′ in which there is evil that is both better than W has more evil than W.

In W′, for instance, each person might suffer one unit of subnoticeable (or just noticeable, if its insisted) pain, and also enjoy equal amounts of happiness, and W′ might be better overall than W. For much more on these issues in a different context, see Parfit (1984, Chapters 17–19) and Temkin (1993).

21. Is it possible for all agents to violate all requirements of justice? It is difficult to see why not, though we might be able to conjure up a case where S's violation of moral requirement R at time t is one that S′ could not simultaneously violate. But I'm placing no such restrictions on violating all of the requirements of justice.

22. This is not because I'm stipulating that U_E is nonempty, which would yield the result trivially. But I'm also not claiming that such an argument could not be cogently advanced here.

23. Aylish Chantler and Klaas Kraay urge that there need not be a world in which every member of a morally perfect world always goes wrong. There need only be several worlds in which each sometimes goes wrong. There in fact needs to be both. There are various ways in which moral agents can go wrong, individually and collectively, and each of these is prohibited by the requirements of justice. The failure to go wrong in any of these ways is what makes a morally perfect world so valuable.

24. This might be construed as an incipient, very weak version of PAP. The alternate possibilities in this case need not be options that an agent is *able* to bring about. They need only be options that are not metaphysically impossible. There is more discussion of this point below.

25. See David Lewis (1981, 116–17). Compare Lewis's illustration of strong compatibilism:

> Had I raised my hand, a law would have been broken beforehand. The course of events would have diverged from the actual course of events a little while before I raised my hand, and at the point of divergence there would have been a law-breaking event—a divergence miracle, as I have called it. But this divergence miracle would not have been caused by my raising my hand . . . Nor would it have been caused by any other act of mine, earlier or later. (116–17)

26. See Plantinga (1974, 165ff.). Plantinga suggests a version of PSF that indexes the ability to act at distinct times. There are interesting questions concerning the conditions under which someone could have done otherwise or someone was able to do otherwise. Among the well-known sources, see Lehrer (1976), but see also Campbell (1997).

27. Note that PGP does not state that S is able to actualize a world in which S does other than A. PGP does not even require that a world in which S does other than A is nomologically possible. It requires minimally that such a world be metaphysically possible.

28. The relevant sort of control for weak compatibilists is typically guidance control. See especially Fischer (2012) and also O'Connor (2010).

29. Would God not allow a universe to get as bad as UE0? First, I'm certain that God cannot decide which worlds are possible by allowing or disallowing them. Worlds are necessarily existing objects, if they exist at all, so exist whether they are allowed or not. But second, there must be a universe UE0 if there is a universe UB0. The existence of the latter entails the existence of the former. That is a central argument in the essay.

30. Thanks to Klaas Kraay and Don Turner for many insightful comments that really improved the paper. I would also like to thank Don Page, Rich Davis, and members of the audience at the God and the Multiverse Workshop (Ryerson University, February 15–16, 2013) for extremely useful questions and comments.

REFERENCES

Almeida, M. 2012. *Freedom, God, and Worlds.* Oxford: Oxford University Press.

Bricker, P. 2001. "Island Universes and the Analysis of Modality." In *Reality and Humean Supervenience: Essays on the Philosophy of David Lewis,* edited by G. Preyer and F. Siebelt, 27–55. Lanham, MD: Rowman and Littlefield.

Campbell, J. K. 1997. "A Compatibilist Theory of Alternative Possibilities." *Philosophical Studies* 88: 319–30.

Cohen, G. A. 2008. *Rescuing Justice and Equality.* Cambridge, MA: Harvard University Press.

Fischer, J. M. 2012. *Deep Control.* Oxford: Oxford University Press.

Frankfurt, H. G. 1969. "Alternate Possibilities and Moral Responsibility." *The Journal of Philosophy* 66: 829–39.

Gauthier, D. 1987. *Morals by Agreement.* Oxford: Oxford University Press.

Hume, D. 1981. *A Treatise of Human Nature.* 2nd ed. Edited by L. A. Selby-Bigge. Oxford: Clarendon Press.

Kant, I. 1953. *Groundwork of the Metaphysics of Morals.* Translated by H. J. Paton. New York: Harper and Row.

Lehrer, K. 1976. "'Can' in Theory and Practice: A Possible Worlds Analysis." In *Action Theory,* edited by M. Brand and D. Walton, 241–70. Dordrecht: D. Reidel.

Lewis, D. K. 1981. "Are We Free to Break Laws?" *Theoria* 47: 113–21.

Lewis, D. K. 1986. *On the Plurality of Worlds.* Oxford: Blackwell.

Mackie, J. L. 1955. "Evil and Omnipotence." *Mind* 64: 200–212.

Nozick, R. 1975. *Anarchy, State, and Utopia.* New York: Basic Books.

O'Connor, T. 2010. "Free Will." In *The Stanford Encyclopedia of Philosophy,* edited by E. N. Zalta. http://plato.stanford.edu/entries/freewill/.

Parfit, D. 1984. *Reasons and Persons.* Oxford: Oxford University Press.

Plantinga, A. 1974. *The Nature of Necessity.* Oxford: Oxford University Press.

Rawls, J. 1975. *A Theory of Justice.* Cambridge, MA: Harvard University Press.

Rowe, W. 2002. "Can God Be Free?" *Faith and Philosophy* 19: 405–24.

Temkin, L. 1993. *Inequality.* Oxford: Oxford University Press.

Van Inwagen, P. 1980. "Indexicality and Actuality." *Philosophical Review* 89: 403–26.

8   On Multiverses and Infinite Numbers

Jeremy Gwiazda

A multiverse comprises many universes, which quickly leads to the question, How many universes? There are either finitely many or infinitely many universes. The purpose of this essay is to discuss two conceptions of infinite number and their relationship to multiverses. The first conception is the standard Cantorian view. But recent work has suggested a second conception of infinite number, on which infinite numbers behave very much like finite numbers. I will argue that that this second conception of infinite number is the correct one and analyze what this means for multiverses.

When it comes to infinite number, the overarching question that I will address is this: which objects are the infinite numbers? How do I mean this question? In the finite case, you have the finite whole numbers, also called the natural numbers, numbers like 7, 13, and 106. I am interested in this question: how should these numbers, the finite natural numbers, be extended into the infinite?[1] I will first address this question and then discuss how that conception of infinite number bears on multiverses. As a preview, if you ask the question "How many universes are there?" then I think that any sensible finite answer is constrained to a finite natural number. You have to say that there are 7 universes, or that there are 106 universes. It doesn't make any sense to say that there are 2.5 universes, or that there are –2 universes, or that there are $3 + 2i$ universes. Now, similarly, I am going to argue that if you give an infinite answer to the question "How many universes are there?" then any sensible infinite answer is constrained to an infinite natural number. So that is why I think it is important to answer this question: which objects are the infinite natural numbers?

That is a bit of an overview. Let us turn to the concept of infinite number. I am going to tell a bit of a story. This is a story in which the whole numbers were not moved into the infinite in the correct way. The standard view of infinite number is a Cantorian view, and I am going to tell a story, involving concrete objects, that tries to highlight the sort of mistake that I think was made. Consider the following items (and concepts): male birds, female birds, and black cars in the United States. Somebody comes along, and he (his concepts) carves up the world as follows. When he sees a male bird, he says, "There is a male bird". But he thinks that the black cars in the United States are the female birds. So when he sees a black car in the

Table 8.1

Object	Concept/Calls Them
Male Birds	"Male Birds"
Female Birds	
Black Cars in the United States	"Female Birds" [wrong conception]

United States, he says, "There is a female bird". Table 8.1 summarizes this.

Now, I think that this person would run into all sorts of questions and problems. For example, Why can't male birds and female birds produce viable offspring? Why are female birds so large? Why, when female birds cross the border into Canada, do they disappear? There would be a whole host of questions that would arise. I suggest that it would be unwise to tackle these problems head-on. Instead, we should try to change this person's way of carving up the world.

I am just going to assume that there are better and worse ways for our concepts to carve up the world. I won't argue for that point. I do just think that that is true, and I suggest that the situation described previously is not a good way to carve up the world. Now the question becomes: is there any way of resolving this dispute, that is, is there any way of straightening this person out? I suggest that there is. We agree on male birds, that is, both I and this confused person point to a male bird and say, "There is a male bird". The disagreement is that I think that the female birds are the female birds, whereas this person thinks that the black cars in the United States are the female birds. Since we agree on male birds, the way to go is to tell this person, "Hey, the female birds are a lot more like the male birds than are black cars in the United States". I think that this is the way out of the confusion. You begin from your point of agreement, to argue that the female birds are the female birds. This argument isn't profoundly deep and subtle. It is simply based on the idea that the more something walks like a duck and quacks like a duck, the more likely it is to be a duck.

Now let's turn to numbers. I do think that Cantor was trying to move the whole numbers into the infinite. Consider this quotation from Cantor (as quoted in Hallet 1984):

I arrive at a natural extension, a continuation of the sequence of real, whole numbers which leads me successively and with the greatest security to the increasing powers [here he means the infinite ordinals] whose precise definition has failed me until today. (60)

Here is another:

This I do in my transfinite number theory . . . with a definition based on the general concept of well-ordered set. Here I will show, in the clearest possible way, that we are here concerned with real concrete numbers in the same sense as 1, 2, 3, 4, 5, . . . that have been designated and looked upon as numbers from ancient times. (51)

So I do think that Cantor was trying to move the whole numbers into the infinite.[2]

What I am now going to suggest is that the Cantorian infinite (the infinite ordinals and cardinals) are not the infinite numbers. Table 8.2 is meant to mirror Table 8.1, but it involves finite numbers, infinite numbers (and here I mean infinite natural numbers in a nonstandard model of the reals), and the Cantorian infinite.

Table 8.2

Object	Example	Concept/Calls Them
Finite Numbers	I I I I I	"Finite Number"
Infinite Numbers	I I I . . . . . . I I I I I . . . . . . I I I	
Cantorian Infinite	I I I I I . . .	"Infinite Number" [wrong conception]

In Table 8.2, finite numbers, infinite numbers, and the Cantorian infinite[3] are playing the roles, respectively, of male birds, female birds, and black cars in the United States. So that when someone says, pointing at ω, "there is an infinite number", that is exactly the same sort of mistake as when someone points at a black car in the United States and says "there is a female bird".

What is the argument here? Essentially it boils down to the claim that the infinite numbers in a nonstandard model of the reals just behave, and are much more like, the finite numbers, than the Cantorian infinite. So again, it's the same idea that if we agree on male birds, I'll just point out that the female birds are more like the male birds than are black cars in the United States. Similarly, the infinite numbers in nonstandard model just behave much more like the finite numbers than does any example of the Cantorian infinite.[4]

Many people have not run across these nonstandard numbers, and so let us consider some examples of this similarity of structure and behavior. Any finite number, when you consider order, has a first element with a successor, a last element with a predecessor, and middle elements with both of these. Infinite numbers share this structure (see Table 8.2, where I use strokes and dots to represent an infinite number). There is a first element that has a successor, a last element with a predecessor, and middle elements that have both predecessors and successors. Any infinite number (in a nonstandard model of the reals, which I sometimes drop for brevity) is even or odd. Any infinite number is prime or composite. If you add 1 to any infinite number, you get a larger number; it is larger by 1. So I suggest that when we ask this overarching question—Which objects are the infinite numbers? or even, What does an infinite number look like?—the infinite numbers are infinite numbers in nonstandard model of the reals. They simply behave and look very much like finite numbers.[5]

Abraham Robinson developed these numbers in the 1960s. He writes:

> In the fall of 1960 it occurred to me that the concepts and methods of contemporary Mathematical Logic are capable of providing a suitable framework for the development of Differential and Integral Calculus by means of infinitely small and infinitely large numbers. (Robinson 1996, xiii)

Recall that the overarching question we are considering is this: which objects are the infinite numbers? Another argument (that the infinite numbers in a nonstandard model of the reals are the infinite numbers) can be made simply by considering language. For example, Robison writes (1996, 51), "Thus any finite natural number is smaller than any infinite natural number". To my knowledge, no one has ever been inclined to call any example of the Cantorian infinite an infinite natural number. So I think that argument can also just come directly out of language.[6]

Now let me say a bit about the importance of this conception of infinite number outside of the context of multiverses, because I think there might be the concern that this is just a verbal issue. A person might come along and say, "That's OK. I'll agree to call the infinite numbers in a nonstandard model 'the infinite natural numbers'. I'll call the Cantorian infinite 'the Cantorian infinite'. So what's the problem?" Well, in my initial example with birds and cars, these questions that arose could be tackled head-on, the sort of silly questions I raised. But I think a better response to that person is to say, "You are just not carving up the world in the right way; you aren't talking about female birds". Similarly, in the case of infinite number, I think that certain issues arise, for example, with paradoxes of the infinite in philosophy or the continuum hypothesis in mathematics, where I don't think the issues should be tackled head-on. I think the correct response is to say, "You aren't talking about infinite numbers; you aren't carving up the world the right way".

Here are some examples. Consider Thompson's lamp—a lamp that begins on, after which the button is pressed to toggle the lamp on and off at times one-half, three-quarters, seven-eighths, fifteen-sixteenths, and so on (Thompson 1954). The button presses are happening more and more quickly, each within half the time of the previous push. Then the structure of the button presses is the structure of ω, that is, the structure of the positive integers. There is a first, a second, a third, a fourth button press, and so on. Thompson tried to create a problem by asking, What is the state of the lamp after these infinitely many button presses? Many people think Benacerraf (1962) successfully replied to Thompson by saying that the state of the lamp is not determined at two minutes. But I think the correct response is to say, "That's not infinitely many, that's not an infinite number of button presses. If you are talking about infinitely many, you have to be talking about an infinite natural number in a nonstandard model of the reals". Note that any such number has to be even or odd. So if the button were pressed an even infinite number of times, then the lamp would be in its starting state; if the button were pressed an odd infinite number of times, then the lamp would be in the opposite of its starting state. When you actually talk about infinite numbers, not only is there no paradox, there is not even any lingering confusion.[7]

Another example is a Zeno sphere. This would be an object centered at an origin that has shells of increasing radius. There is a spherical shell of radius one-half, a shell of radius three-quarters, a shell of radius seven-eighths, and so on. These shells go on approaching 1, getting closer and closer and closer but never getting to 1. The shells get increasingly thin so that no two shells touch each other. Note that there is no outermost shell. People ask questions like, What happens when two Zeno spheres collide? What happens when you shine a light on a Zeno sphere? I tend to think that the literature on this topic is a bit tortuous.[8] I also think the correct response here is to say, "That is not an infinite number of shells. That's not a possible thing you are describing". I don't think that ω exists in any actual determined sense, so there is a very Aristotelian feel to this view. If you want to talk about an object with an infinite number of shells, then you need to be talking about an object that has an infinite number of shells. And then there will be an outer shell, and there is no problem with these things colliding or with light bouncing off them. Again, no paradox or puzzle remains.

A third example is the spaceship paradox. A spaceship travels one mile in half an hour, another mile in a quarter of an hour, a mile in an eighth of an hour, a mile in a sixteenth of an hour, and so on. So it keeps travelling miles but faster and faster and faster. You can ask the question: after these infinitely many trips, where is the spaceship? There is no good answer here. But again I think that the correct response here is that you are not talking about infinitely many; ω is not an infinite number. If you do talk about an infinite number, it has to be an infinite number. Call it M. Then if you make infinitely many, M many, trips of one mile, then there is no problem—you are M miles away. So I do think that you dissolve a number of, or maybe all, paradoxes of the infinite by recognizing the correct conception of infinite number.[9]

To take stock, when you are talking about infinitely many, it has to be an infinite integer. Any infinite integer has a certain structure, including having a last element. You get rid of many problems and puzzles of the infinite.[10] Now perhaps it is an overstatement to write, as I just did, that when talking about infinitely many, "it has to be an infinite integer" involved. A person can carve up the world so that the black cars in the United States are called and conceived of as the 'female birds'. But confusion ensues. Similarly, correctly identifying the infinite numbers need not be done, but it should be done. Arriving at the correct conception of the infinite numbers leads us out of a great deal of confusion. What are the costs? I see none; ω can still be used, discussed, and explored. Nothing is lost. The Cantorian confusion is to think that ω is an infinite number and that it is an actual, determined, static entity, so that, for example, a sphere can have ω shells. Rather, ω is merely potentially infinite. There is no benefit in wracking one's brain in attempting to determine what happens when two Zeno spheres collide, because a Zeno sphere cannot exist.[11] By contrast, it makes perfect sense to use ω to talk about limit processes; certainly the sequence 1/1, 1/2, 1/3, 1/4, 1/5, . . . has a limit of 0. No legitimate uses of ω are barred.[12]

How does this conception of infinite number bear on multiverses? One answer is that this position has implications for the mathematics of multiverses. In particular, if there are infinitely many universes, then there is some infinite (natural) number of universes. The mathematics then becomes very like the finite case. It's almost exactly the same. And so if M is an infinite integer, and you have M universes of value 1, then the total value is M times 1, or M. If you then add in another universe of value 1, the total value is greater, namely, M + 1. Adding a good universe to a multiverse always increases the total value.

Adding a good universe may, however, decrease the average value. If you have M universes of value 1, then the average value is 1. Let's say that you then add a universe of value 0.5. The average value is now lower, it is infinitesimally less than 1, namely, $(M + 0.5)/(M + 1)$. The average value has gone down.[13]

Let us consider one more example to highlight the fact that the mathematics of multiverses, on what I claim is the correct conception of infinite number, is very much like the finite case. Imagine that you have universes numbered 1, 2, 3, . . . and so on through M – 2, M – 1, M. You can ask: what is the total value? Now if M is a finite number, the answer is given by the formula M(M + 1)/2. So if M is 4, and so there are universes of value 1, 2, 3, and 4, then the formula gives a correct total value of $4*5/2 = 10$. And to see that this is correct, note that $1 + 2 + 3 + 4 = 10$. The same formula holds in the infinite case. If M is an infinite integer, you have universes valued 1, 2, 3, . . . , M – 2, M – 1, M, and you ask for the total, then it is still given by that formula M(M + 1)/2. If you ask for the average value, this case is again like the finite case, with the average value given by (M + 1)/2. So again, the summary is that the mathematics becomes very much like the finite case. The mathematics of multiverses is greatly simplified on the correct conception of infinite number. With the math, just like the finite case, it is possible to talk about total values, average values, and so forth. In the Cantorian infinite, matters are not so clear.

Now, what does this mean for multiverses and theism? Certainly that depends on the specific theistic position under consideration and also on the multiverse theory that is proposed. For example, here is a theory of a multiverse: The multiverse begins with one universe and then splits into some finite number of universes. At any point in time, a single universe can split into a finite number of universes. The conception of infinite number that I have proposed does not have any bearing on this type of multiverse, because at any given time there are only finitely many universes. For example, there may be one universe, then two, then four, then eight, and so on. The number of universes doubles and therefore grows without bound, but at no particular time do an infinite number of universes exist.

On other theories of multiverses, an infinite number of universes exist. On these theories, I claim, the number of universes in a multiverse must be an infinite number. Let me now discuss how the conception of infinite number outlined earlier relates to this sort of multiverse and theism. Our universe is a rather remarkable thing. The vastness of the universe, the solar system, Earth, humans, morality, beauty, order—these are all things that seem to cry out for an explanation. Many explanations are possible.[14] Some theists argue for God's existence (or at least for the existence of a designer) from the fine-tuning of the universe. For example, it is often pointed out that if physical constants had been slightly different, then humans could not have existed. The universe is fine-tuned for life, which may raise the probability that the universe was designed to be able to support life. An atheist might then reply that there is a multiverse, and so it is not surprising that we find ourselves in one of the universes able to support the existence of human life. And indeed, if there are infinitely many universes, then the claim that fine-tuning provides support for the existence of God is undercut. Here I think we are in the realm of probability, where we can ask the question: what is more likely, the existence of God or the existence of infinitely many universes? Though I won't argue for these claims here, I believe that smaller numbers are simpler than larger numbers and that simplicity raises the (prior) probability of a hypothesis. Richard Swinburne (2004) has argued for these two points. It follows that it is most likely that one universe exists.

The atheist in this dialogue can attempt to argue that the infinite is simple, because the infinite is a sort of endpoint or maximal entity. This move attempts to raise the probability of a multiverse that contains infinitely many universes. However, the Cantorian position on infinite number may seem simple, but if my view of infinite number is correct, then a specific infinite number itself is not simple—any infinite number is mysterious in its specificity. This then detracts from the (prior) probability that there is a multiverse. Any infinite number is just as mysterious and specific as a finite number. By this I mean that if you say that there are 56 universes, questions arise. Why not 57? Why not 55? Why not any other number? Why an even number? All these questions also arise when you say that there are an infinite number of universes. If there are M universes, where M is an infinite integer, then M is even or odd. Questions arise: Why not M + 1 universes? Why not M − 1? Why not any other number? If M is even, why an even number? And so I think that a problematic feature of this position on infinite number for multiverses is that you lose this appeal to simplicity, which then affects prior probabilities. Thus, if we focus on multiverses versus a designer as hypotheses explaining order and fine-tuning, this view of infinite number makes the multiverse less likely and therefore makes the existence of a designer correspondingly more likely. The overriding point is that the hypothesis of one universe is far more likely than the hypothesis of any larger specific number of universes, whether finite or infinite. And so this position on infinite number, I believe, provides some support for theism and the existence of one universe.

Though earlier I discussed two competing explanations for the existence of our universe, I certainly did not mean to imply that the list was exhaustive, nor did I mean to imply that theism and a multiverse are contradictory. For example, perhaps God exists and creates all and only those universes that are worthy of creation.[15] I think that the considerations outlined previously lower the probability of there being a multiverse (relative to the probability of there being one universe), as any infinite number is not simple. Ultimately, strong reasons would have to be presented to have a successful argument for the existence of a multiverse. I think that it is currently an open question whether there are any such strong reasons. The numerical considerations presented make it most likely that one universe exists—ours.

Before summarizing and concluding, I tie up a few points and try to anticipate a few objections. One key point: I don't want to overstate the claim that I am making. I said earlier that I would tell a story in which the whole numbers were not moved into the infinite correctly, in which Cantor did not undertake the task correctly. And indeed I believe that the infinite numbers have been misidentified. When you say "there is an infinite number" or are talking about "infinitely many", you should be referring to an infinite integer. But I don't mean to detract from Cantor's contribution. I am not claiming that ω is completely useless. Nothing of the sort. My claim is a limited one about the correct conception of infinite number. ω is certainly infinite, insofar as it is not finite, but it is not an infinite number. I think that ω is a potential infinity, and so, for example, of course you can talk about infinite sums. The summation of $1/2^i$, for i ranging from 0 to infinity, equals 2. This is a true, meaningful, and important claim. But I think that it is also important to keep in mind the underlying meaning. The underlying meaning is that for any epsilon, no matter how small, I can tell you where, from some point on, the partial sums are within epsilon of 2. It is not somehow that you really get to 2. This fact provides evidence that ω is a potential infinity.[16] This essay, in this regard, and as I have mentioned, has an Aristotelian feel. In contrast, when you take a hyperfinite sum, so that you add up infinitely many things, where the sum runs from 1 to M and M is an infinite integer, I think that you are dealing with an actual infinity.[17] And your sum really does sum to a specific number. The sum from i = 1 to M of $1/M$ really sums to 1. The sum from 1 to M of $1/2^i$ is actually infinitesimally shy of 1, so there is no sort of potential infinity in these cases involving the correct infinite numbers. This bolsters the claim that these objects really are the infinite numbers.

But why, an objector may ask, reject other conceptions of infinite number? The reason is that I am asking a specific question, and I think that there is a single correct answer. In particular, the infinite numbers, correctly identified, are the infinite numbers in a nonstandard model of the reals. By way of comparison, imagine that there were 120 years of confusion on the bird–car issue, so that people see a black car in the United States and say, "There is a female bird". I think that this is simply wrong. The female birds are the female birds. Similarly, the infinite natural numbers should be properly identified. But why, the objector may continue, is there only one correct conception of infinite number? And why must the finite numbers be our guide to determining what the infinite numbers are? To the second question I simply ask: what else do we have? If there are any other suggestions as to how to arrive at the infinite numbers, let them be produced.[18] In reply to the first question, note that the person who thinks that black cars are the female birds can ask the same question, namely: why is there only one correct conception of female bird? Why not have black cars in the United States and female birds fall under the concept of female bird? Whatever response is given here can be used in the case of infinite number. The response, at its core, is simply that this way of breaking the world into concepts is not a good one. It most likely will lead to confusion. In the context of infinite number, I have suggested two examples of such confusion (which arises from thinking of the Cantorian infinite as being one example of infinite numbers): philosophical paradoxes involving the Cantorian infinite and the continuum hypothesis.

Another potential objection is this: what if the universe really does go on infinitely far in all directions? And here my response is that I don't think that this is describing a genuine possibility. Just as ω is not an actual infinite number, so, too, a ray is not any actually determined distance. If you want an infinite universe, then it has to be contained within an actually infinite distance, and any actually infinite distance has endpoints. You can then embed open-ended space within that, in a copy of the hyperreals (in one dimension). A very similar objection, returning to the context of number, is this: what if there just really is a multiverse that has universes numbered 1, 2, 3, . . . and so on? But here again my response is that I don't think that that is describing anything. There is no set of all finite numbers; ω is not an actual determined thing, and so I don't think that that is a real possibility.[19]

One final objection: Why does there have to be a number of universes? Someone may say that there not only doesn't have to be a number of universes but furthermore that the universes don't even need to form a set. Here my reply is that I think that the finite has to be our guide. And so I mentioned before that any finite answer (to the question "How many universes are there?") is constrained to a finite number. Similarly, any sensible infinite answer is constrained to an infinite number.

To summarize and conclude, there are better and worse ways for our concepts to carve up the world. I do not believe that the infinite numbers have been properly identified. I think that the infinite numbers are as I've described. Sensible infinite answers to the question—how many?—must be infinite numbers. This view of infinite numbers simplifies the mathematics of multiverses but detracts from the probability of there being any specific number of universes in a multiverse. In the context of the competing explanations of God versus a multiverse to explain fine-tuning, this lowers the probability of a multiverse and may thereby bolster the probability of theism insofar as the existence of only one universe is more likely than the existence of many universes.

NOTES

1. We might also ask these questions: What does one infinite number look like? And in what sort of system does it operate? These questions are addressed in Gwiazda (2011, 2012).
2. Perhaps, though, this is overstated. It might be more accurate to say that Cantor was concerned with a particular mathematical problem, which then led him to the ordinals. He then thought that these ordinals were the correct extension of the finite natural numbers into the infinite. It is this last point that is important—Cantor thought that his ordinals answered this question: what are the infinite (natural, whole) numbers?
3. I should say a bit about how I think of these numbers. I really do just think of them in terms of strokes and dots. And so 5, earlier, is an example of a finite number—it is five strokes. It does have a certain structure. So, for example, order is involved, so in 5, or | | | | |, there is a first element that has a successor, a last element that has a predecessor, and middle elements with both predecessors and successors.
4. I said earlier that the argument was not profoundly deep or subtle. And with all my talk of birds and cars, I worry that it is easy to dismiss my argument. But I am claiming that people, for more than 100 years, have not correctly identified the infinite numbers. And just as misidentifying the female birds would lead to all sorts of confusion, I claim that people have been led into all sorts of confusion based on this misidentification of the infinite numbers. The argument may not be deep or subtle, but I do suggest that the conclusion is correct and important.
5. Just like female birds behave and look very much like male birds.
6. I also think that the same point also applies to infinite distance. I don't think there is any distance to a ray or even a real number line. I don't think those are examples of infinite distance. I think that if you want to talk about infinite distance, it has to be infinite distance between two points, and the structure of this is going to be very similar to the structure of an infinite integer. Just as infinite integers have a certain structure, so, too, infinite distance has this structure: |————————> . . . <————————> . . . <————————|
7. This topic is addressed in Gwiazda (2013b).
8. An actual, determined, static Zeno sphere is physically impossible, I believe. For an example of an argument involving Zeno spheres, which to my mind

does not adequately explain what happens when light shines on such an object, see Peijnenburg and Atkinson (2010).

9. In fact, I think that an interesting challenge is to try to arrive at any paradoxes, while referring to the correct conception of infinite number. That is, I am suggesting that perhaps no paradoxes of the infinite remain.

10. I also think it is potentially important for physics and math to correctly identify the infinite numbers. For example, in cosmology, if the universe is infinite, then it has some infinite distance. Just as any finite distance (in, say, one dimension) is bound by endpoints, so, too, a universe of infinite distance must be bound. Cosmologists should consider the possibility that the universe is infinite and bound. In math, I think that something like the continuum hypothesis is just a nonstarter. I don't think that there is any number of positive integers, and so it is no surprise that when you ask about the positive integers and the powerset of the positive integers, and ask the question, "Are there sizes in the middle?" it is no surprise that you don't get a clean answer when there is not even any number of positive integers.

11. Just as it is most likely not fruitful to explore why a black car and a male bird cannot mate. How our concepts carve up the world matters.

12. Legitimate uses are precisely those where it is recognized that ω is merely a potential infinity.

13. Though perhaps something of a tangent, I do find this an interesting question: which value matters more in the context of multiverses, the average value or the total value? Here I admit that I do not have strong intuitions one way or the other. If Picasso painted a painting that was worth painting, but it brought down the average value of his paintings, it's unclear to me whether this is good or bad, or maybe more detail is needed. Perhaps producing paintings and producing universes is disanalogous, but in neither case do I have a sense of whether the total or the average is the relevant consideration.

14. Of course, there might be no explanation. The existence of the universe might simply be a brute fact.

15. For one argument to this conclusion, see Kraay (2010).

16. This topic is discussed in Gwiazda (2013a).

17. For a test to determine whether a structure is potentially infinite, see Gwiazda (2013a).

18. To push this point a bit further, if someone points at his cat and says, "There is an infinite number," that seems wrong. But if you give up using finite numbers to arrive at the infinite numbers, it is genuinely unclear to me how to argue that this person is wrong. Again the question is, if we give up finite numbers, what do we then use to arrive at the infinite numbers?

19. When it comes to the size of ω, there is no number of natural numbers. What is the argument? I think that two things are the case. The first has been the main theme of this chapter: infinite numbers, properly understood, are infinite numbers in a nonstandard model. Second, I also think that the correct way to judge the relative sizes of infinite numbers is via what is sometimes called SUBSET, where if A is a proper subset of B, then A is smaller than B. So if you add 1, then you have a larger set, larger by 1. This contradicts judging sizes based on BIJECTION, or what is sometimes called Hume's Principle. If you grant these two claims, then it turns out that ω is too large to be a finite number, as any finite number is a proper subset of ω. But ω is also not an infinite number, as ω, structurally, is the initial segment of any infinite number. So ω is this strange middling beast that is too large to be finite in number but too small to be infinite in number. That is why I think that there is no number of natural numbers, and for example,

why there cannot be ω–many universes in a multiverse. See Gwiazda (2011, 2012) for this argument in more detail.

REFERENCES

Benacerraf, P. 1962. "Tasks, Super-Tasks, and the Modern Eleatics." *The Journal of Philosophy* 59: 765–784.

Gwiazda, J. 2011. "Infinite Numbers Are Large Finite Numbers." http://philpapers. org/rec/GWIINA.

Gwiazda, J. 2012. "On Infinite Number and Distance." *Constructivist Foundations* 7: 126–130.

Gwiazda, J. 2013a. "Throwing Darts, Time, and the Infinite." *Erkenntnis* 78: 1–5.

Gwiazda, J. 2013b. "Two Concepts of Completing an Infinite Number of Tasks." *The Reasoner* 7: 69–70.

Hallett, M. 1984. *Cantorian Set Theory and Limitation of Size*. Oxford: Clarendon Press.

Kraay, K. J. 2010. "Theism, Possible Worlds, and the Multiverse." *Philosophical Studies* 147: 355–368.

Peijnenburg, J., and D. Atkinson. 2010. "Lamps, Cubes, Balls and Walls: Zeno Problems and Solutions." *Philosophical Studies* 150: 49–59.

Robinson, A. 1996. *Non-Standard Analysis*. Rev. ed. Princeton, NJ: Princeton University Press.

Swinburne, R. 2004. *The Existence of God*. 2nd ed. Oxford: Oxford University Press.

Thomson, J. F. 1954. "Tasks and Super-Tasks." *Analysis* 15: 1–13.

Pantheistic Multiverses

9 Multiverse Pantheism

Yujin Nagasawa

> Nothing is superior to the world; the inevitable consequence . . .
> is . . . that the world, therefore, is a Deity.
>
> —Cicero

Traditional pantheism holds that the world is identical with God, where the term 'world' normally refers to our universe. Multiverse pantheism, on the other hand, holds that God is identical with the totality of all universes, that is, the multiverse that includes our universe. In this essay, I examine strengths and shortcomings of multiverse pantheism. I argue that multiverse pantheism is attractive insofar as it overcomes many difficulties that traditional pantheism faces. I claim, however, that multiverse pantheism itself faces an unwelcome consequence that is difficult to eliminate.

INTRODUCTION

Pantheism is traditionally formulated as the view that God is identical with the world. This formulation tacitly assumes that the world consists only of our universe.[1] Thus, traditional pantheism can be construed as follows.

Traditional pantheism: God is identical with our universe.

According to some contemporary scientists and philosophers, however, our universe is not alone—there are many more, perhaps infinitely many more, universes. That is, the world consists of or *is* the multiverse, a set of many universes. If the multiverse truly exists, then pantheists might be attracted to the following view instead:

Multiverse pantheism: God is identical with the multiverse.

Multiverse pantheism is an unusual view and, as far as I know, it has not been endorsed explicitly by any philosophers. In this chapter I argue that multiverse pantheism is more attractive than traditional pantheism in many respects, particularly insofar as it is not vulnerable to strong objections to traditional pantheism. I do not, however, intend to offer an unequivocal defense of multiverse pantheism, because multiverse pantheism entails an unwelcome consequence which is difficult to resolve.

This chapter has the following structure. In the next section, I offer an overview of pantheism and its alternatives. In the third section, I discuss three prominent objections to traditional pantheism. In the fourth, I introduce multiverse pantheism. In the fifth, I explain the relationship between multiverse pantheism and modal realism. In the sixth, I argue that multiverse pantheism avoids all the objections to traditional pantheism. In the seventh, I address what I regard as the most difficult problem for multiverse pantheism and argue that there may be no satisfactory solution to this problem. The final section concludes.

PANTHEISM AND ITS ALTERNATIVES

Since there are many definitions of pantheism, the difference between pantheism and its alternatives, such as traditional theism and pan*en*theism, is not always clear. Robert Oakes (1983) argues, for example, that traditional theism might entail pantheism. To take another example, John Culp (2013, Section 2) points out that although many views in the history of philosophy have been "accused of pantheism by their contemporaries, their systems can be identified as panentheistic". I submit, however, that pantheism and its alternatives can be clearly distinguished if we formulate them mereologically. As mentioned earlier, pantheism is traditionally formulated as the view that God is identical with the world. The mereological formulation agrees with this. So, according to this formulation, pantheism entails that the world entirely overlaps with God, and vice versa. Traditional theism, on the contrary, says that God and the world are ontologically distinct and hence do not overlap at all. According to this view, God chose to create a world that is distinct from Himself, and that is our world. Pan*en*theism lies between these two views. It says that the world is a proper part of God but is not an essential part of God. God and the world overlap but the overlap is only partial and not essential. God subsumes the entire world but He is not exhausted by the world. God is, therefore, in an important sense beyond the world.[2]

There are many other ways of formulating pantheism and its alternatives, but in this chapter I adopt the preceding formulation. It entails that traditional pantheism and multiverse pantheism are both versions of pantheism because both of them hold that God is identical with the world. The disagreement between these two views turns only on whether the world consists entirely of one universe—that is, our universe—or the multiverse which includes our universe.[3]

OBJECTIONS TO TRADITIONAL PANTHEISM

Again, traditional pantheism says that God is identical with our universe. This view has attracted a number of objections. The common idea underlying these objections is that our universe cannot be God because it has specific attributes, such as spatiotemporal finitude, evilness, and the lack of worship-worthiness, which are incompatible with our ordinary conception of God. The three objections to pantheism that we discuss here can be construed as alternative ways of elaborating this idea.

Spatiotemporal Finitude of the Universe

The first objection that traditional pantheism faces is that the universe cannot be God because, unlike God, it is finite. Empirical data suggest that the universe began to exist approximately 15 billion years ago and it is, though expanding, finite in size. God is, on the other hand, according to most theists, an infinite being that is free from any spatial or temporal boundaries. The spatiotemporal finitude of the universe, therefore, seems to undermine pantheism. Scientists also suggest that the universe will cease to exist eventually as it collapses into the Big Crunch. If they are right, then the universe is not even potentially infinite in terms of space and time. Furthermore, these observations show that the universe is not a necessary being. The universe is a contingent being like us that begins to exist at one point and ultimately ceases to exist. This seems to undermine pantheists' claim that the universe is God.

The Problem of Evil

The second objection to pantheism says that the universe cannot be God because it is not perfect. God is believed to be a perfect being, so anything that has an evil aspect seems unqualified to be God. Yet the universe includes many evil states of affairs. A similar point has been made against traditional theism. The problem of evil for traditional theism says that an omnipotent and morally perfect God cannot be a creator of the universe because it contains evil states of affairs, which such a being would not permit. In response to the problem, traditional theists often try to shift the responsibility for evil from God to free human agents. They say, for example, that actual instances of evil, such as wars, murders and rapes, do not undermine the presence of God because free human agents, rather than God, are responsible for them. Such a response to the problem of evil is not, however, available for pantheism because, according to pantheism, free human agents *are* parts of God. Unlike traditional theism, pantheism denies that God exists as an autonomous agent ontologically distinct from free human agents, so it cannot shift the responsibility from God to human agents. If pantheism is true, whatever free human agents do is part of God, so evil states of affairs caused by humans cannot be set aside.[4] Of course, the free will response is not the only existing reply to the problem of evil. Some traditional theists claim, for example, that the problem of evil is untenable because the existence of evil is necessary for instantiating good. That is, there cannot be good without evil. Others endorse so-called skeptical theism, according to which the problem of evil is only an epistemic problem. That is, *we* do not understand why evil exists because our knowledge of morality is too limited and, hence, the existence of evil does not undermine the existence of God. Pantheists might try to undercut the problem of evil by following these lines. The problem of evil, however, remains a serious threat to pantheism because pantheists are committed to hold, unlike traditional theists, that every single instance of evil is part of God.

Worshipworthiness

Worshipworthiness is regarded as an essential property of God. Many, if not most, think that any being that is not worshipworthy cannot be God (Findlay 1955; Morris 1984; Swinburne 1981). Traditional theists have provided numerous accounts of the grounding of God's worshipworthiness but the most common are the following two. (i) God is worshipworthy because He created us. God, and only God, is the fundamental basis of our existence and He chose to create us. Therefore, worship is our appropriate response to God. (ii) God is worshipworthy because He is the greatest possible being. According to the Anselmian conception, God is something than which no greater can be thought, which seems to entail that God has many impressive attributes such as knowledge, power and benevolence to the highest degree. That is, there is no being that is greater than God. That is why God, and only God, deserves our worship.[5]

It seems that pantheism cannot show that God is worshipworthy, because the universe fails to conform to the preceding two accounts. First, pantheism does not hold that the universe is a creation of God. It denies that the universe was created by an omnipotent and morally perfect being. Second, it is clear that the universe, that is, the pantheistic God, is not the greatest possible being because there are many other possible beings that are greater than the universe. We can easily imagine, for example, a universe that is identical with our universe except that it is free from certain subtle negative features of our universe. Such a universe is better than our universe, and, hence, our universe is not the best possible universe, let alone the best possible being. Leibniz famously bites the bullet and argues that our universe *is* the greatest possible universe, but most philosophers find such a claim highly counterintuitive. Moreover, it seems obvious that, unlike God, the universe is neither omniscient nor omnipotent nor morally perfect.

In the sixth section, I will argue that multiverse pantheism is free from the preceding objections to traditional pantheism. But before doing so, I will explain multiverse pantheism in detail.

MULTIVERSE PANTHEISM

There are many models of the multiverse, but the model that I focus on in this chapter is the most extreme one. According to this model, the multiverse consists of all universes that can, logically (or metaphysically), be actualized. Moreover, all of these universes are actual. All the universes are causally and spatiotemporally isolated from one another, but they are ontologically on a par. I explain in the fifth section that the multiverse model in question seems to be essentially identical to modal realism, according to which all possible worlds exist to the same extent that the actual world exists.

A version of multiverse pantheism that I have in mind here (henceforth, simply 'multiverse pantheism') says that God is identical, not with our universe alone, but with the aforementioned multiverse, which includes all universes that can logically (or metaphysically) be actualized and which, of course, includes our universe. Multiverse pantheism appears to differ radically from traditional theism. Surprisingly enough, however, the multiverse pantheistic concept of God shares a number of distinctive features with the traditional concept of God.

First, the multiverse pantheistic concept of God is, like the traditional concept of God, based on the Anselmian thesis that God is the being than which no greater can be thought, although based on a different interpretation of that thesis. According to the multiverse pantheistic interpretation, greatness in the definition should be understood not in terms of the degree to which a being has great-making properties, such as knowledge, power and benevolence, but in terms of the scope of that which it encompasses. By encompassment, I mean inclusion of an entity as part of one's own being. According to this interpretation, the more encompassing a being is, the higher the greatness of that being is. Thus, God, as that than which no greater can be thought, is the maximally encompassing being. And multiverse pantheism says that such a being is the totality of all possible universes that are actualized because there is nothing that can be actualized beyond that totality.

Second, it can be argued that the multiverse pantheistic God shares, at least to some degree, several properties with the traditional God.[6] Such properties include omniscience, omnipotence, omnibenevolence, immutability, eternity, omnipresence, independence, unsurpassability, necessary existence and the property of being a cause of the universe. Again, multiverse pantheism says that God is the totality of all possible universes, a being that encompasses all possible states of affairs and, moreover, that all such possible universes are actual. This means that all possible forms of knowledge, power and benevolence are actual and encompassed by God.[7] This implies that, at least in one sense, the multiverse pantheistic God is omniscient, omnipotent and omnibenevolent.[8] The multiverse pantheistic God is also omnipresent and eternal in the sense that it encompasses all spatiotemporal locations in all possible universes. The multiverse pantheistic God is also independent because, as the totality of all possible universes, its existence does not rely ontologically on any other existence. In fact, there is nothing external to God as there is nothing actual outside the totality of all possible universes. The multiverse pantheistic God is also unsurpassable since there cannot be anything actual that is greater than the totality of all possible universes. The totality of all possible universes is the most encompassing being that can be actualized. I will explain in the next two sections why the multiverse pantheistic God can also be considered a necessarily existent being and a cause of the universe.

Perhaps the only prominent attribute of the traditional theistic God that the multiverse pantheistic God lacks is the property of being an agent with free will. Traditional theism says that God is a person or a free agent with intention. Multiverse pantheism does not attribute such a property to God even though it allows God to subsume all possible persons. Nonetheless, as we have seen, the multiverse pantheistic God shares, at least in some sense, many properties with the traditional theistic God. It seems reasonable, therefore, to view multiverse pantheism as a form of theism, although ultimately, of course, it is a matter of definition whether or not any given view can be construed as a form of theism.

MODAL REALISM AND MULTIVERSE PANTHEISM

Modal realism says that all possible worlds exist to the same extent that the actual world does and that all possible worlds, including the actual world, are ontologically on a par (Lewis 1986). Is modal realism essentially the same as the multiverse model on which multiverse pantheism is based?

On the face of it, there are several differences between these two. (i) While modal realism is concerned with possible *worlds,* the multiverse in question is concerned with possible *universes.* (ii) The multiverse model claims that there are universes that are causally and spatiotemporally isolated from one another. Arguably, this claim is incompatible with modal realism. Bigelow and Pargetter (1987) argue that modal realism cannot allow 'island universes', universes that exist in the actual world but are causally and spatiotemporally isolated. The most prominent proponent of modal realism, David Lewis (1986, 72), himself says, "I cannot give you disconnected space-times within a single world". (iii) Modal realism denies that all possible worlds are *actual;* it says only that all possible worlds *exist* to the extent that our world, that is, the actual world, does. According to modal realism, actuality is merely indexical. There is no ontological difference between the actual world and other possible worlds. The actual world is special for us merely because it is *our* world, just as some other possible world is special for the inhabitants of that world. All possible worlds are ontologically on a par although they are causally isolated from one another. Hence, modal realism says that what distinguishes our world from other worlds is merely indexical. On the other hand, the multiverse model in question says that all possible universes are indeed actual. The multiverse is actual and it actually includes all possible universes.

However, I submit, the differences between modal realism and the multiverse model are merely terminological. First, what modal realism calls 'worlds' are essentially the same as what the multiverse model calls 'universes'. Universes, according to the multiverse model, are spatiotemporally isolated realms which jointly exhaust all possible states of affairs, so they are equivalent to possible worlds according to modal realism. Modal realism *is* incompatible with island universes but what the multiverse model really assumes are not island universes in a single possible world but 'island possible worlds', which are compatible with modal realism. Second, the difference in the treatment of actuality in modal realism and in the multiverse model is also a terminological matter. Modal realism denies that all possible worlds are actual, while the multiverse model holds that all universes are actual. However, what Lewis really means by 'the actual world' is only '*this* world in which we live'. He writes, "Ours is the actual world; the rest are not actual. Why so?—I take it to be a trivial matter of meaning. I use the word 'actual' to mean the same as 'this-worldly'" (Lewis 1986, 92). Given Lewis's definition of actuality, modal pantheists would agree that there is only one actual universe, which is *our* universe and other universes are nonactual. So there is no essential difference between modal realism and the multiverse model. What the multiverse model calls our universe is what Lewis calls the actual world and what the multiverse model calls other possible universes are what Lewis calls other possible worlds.[9]

If the multiverse model is essentially identical to modal realism, there seems to be a version of the ontological argument for multiverse pantheism that is comparable to the version of the ontological argument for traditional theism: again, the being than which no greater can be thought is, according to multiverse pantheism, the totality of all possible universes. Suppose, for the sake of argument, that some of the universes in the multiverse in question, that is, the totality of all possible universes, are not actual (or not existent in Lewis's terminology). It then follows that another totality can be thought that is greater than the totality of all possible universes. Such a totality is thought to encompass all possible universes and all of these universes are thought to be actual (or existent in Lewis's terminology). It is contradictory, however, to say that a being can be thought that is greater than the totality of all possible universes because the totality of all possible universes is that than which no greater can be thought. Hence, it is impossible that some of the universes in the totality of all possible universes are not actual. Therefore, all possible universes in the totality of all possible universes actually exist. Therefore, multiverse pantheism is true.

If this version of the ontological argument is sound, God, as the totality of all possible universes, is also a necessary, self-existent being. A necessary being is commonly understood as a self-existent being that does not require any external cause of its existence. The totality here is assumed to encompass everything, including our universe.

HOW MULTIVERSE PANTHEISM AVOIDS OBJECTIONS TO TRADITIONAL PANTHEISM

We are now ready to consider the strengths of multiverse pantheism. In this section, I argue that, all other things being equal or held constant, multiverse pantheism is more attractive than traditional pantheism because it is not vulnerable to the objections to traditional pantheism which we discussed earlier.

Spatiotemporal Finitude of the Universe

Again, traditional pantheism says that God is identical with our universe. This entails the unwelcome consequence that God is spatiotemporally finite. Multiverse pantheism does not face this problem. Multiverse pantheism identifies God with a set of all possible universes. As the most encompassing being, the multiverse pantheistic God encompasses all possible universes, such that all possible universes that were/are/will be actualizable were/are/will be actualized. Hence, if there are infinitely many universes, there is no beginning or end to the multiverse on the whole. It is true that *our* universe is temporary and finite; it began to exist as the Big Bang took place and it will eventually collapse into the Big Crunch. But this does not mean that the multiverse overall is temporally infinite. Similarly, the multiverse in question does not entail the unwelcome thesis that the universe is spatially finite. If it is possible for any universe to be spatially infinite, the multiverse subsumes such a universe as it subsumes all possible universes that can be actualized. Even if it is impossible for any universe to be spatially infinite, multiverse pantheism can still be considered spatially transcendental, as it entails that for any universe one picks, there is another universe that is spatially distinct. Multiverse pantheism, therefore, avoids the first objection to traditional pantheism.

The Problem of Evil

As you recall, the problem of evil for traditional pantheism says that the universe cannot be God, since it contains evil states of affairs. Multiverse pantheism admits that evil is part of God but insists that this does not undermine the view. On the contrary, it says, given that God is the most encompassing being, He has to subsume all possible states of affairs, including evil states of affairs.

One might wonder, however, why there has to be evil in *our* universe. If God encompasses all possible universes there should be a universe totally free from evil. Why can our universe not be such a universe? Multiverse pantheism responds to this question by saying that there is no reason that our universe, instead of some other universe, has to be free from evil. Multiverse pantheism does not privilege our universe over other universes.

Our universe is one of infinitely many universes constituting the multiverse. Our universe appears morally or metaphysically special to us merely because *we* happen to exist in this universe. In other universes similar to ours there are our counterparts. If we do not suffer from evil, then certain of our counterparts in some other universe, who are as real as we are, have to suffer from evil. Lewis (1986, 127) says that modal realism entails that the net amount of evil in the totality of all possible worlds cannot be increased or diminished. Similarly, the net amount of evil in the multiverse in question cannot be increased or diminished. Given that we and our counterparts are morally equivalent, there is no reason for them to suffer instead of us, and vice versa. Perhaps, as Almeida (2011, 9) says, this is comparable to a situation in which a rescuer can save each of two drowning children but not both.[10]

One might think, however, that multiverse pantheism faces a different form of the problem of evil. The multiverse encompasses all possible states of affairs, including utterly evil states of affairs, such as those in which innocent people are tortured for a long time, possibly an infinite amount of time, without any reason. How could we reconcile this with the idea that God is identical with the multiverse? I address this issue in detail in the seventh section.

Worshipworthiness

Arguably, God is, by definition, worshipworthy. However, as you recall, the third objection says that the traditional pantheistic God cannot be worshipworthy because He is neither a creator of the world nor the greatest possible being. God is simply the universe which we inhabit. Multiverse pantheism does not face this objection. Consider, first, God as a creator of the universe. As I mentioned earlier, the multiverse pantheistic God can be construed as a necessary, self-existent being. This implies that our universe, which is subsumed by the multiverse, can be understood as a being caused by God, the self-existent multiverse. Hence, in one sense, we can say that God is a cause, if not the creator, of our universe. I mentioned earlier that the multiverse pantheistic God is not an agent, so multiverse pantheists cannot say that God is the person who freely chose to create the universe. Yet there is a sense in which God is a creator or cause of the universe and the fundamental basis of our existence.

Consider, then, God as the greatest possible being. Traditional pantheism says that God is our universe, but clearly our universe is not the greatest possible being. Multiverse pantheism, on the other hand, says that God is the multiverse. Again, the multiverse in question encompasses all possible universes, so given that we view greatness as a matter of encompassment rather than of great-making properties, the multiverse pantheistic God *is* the greatest possible being. There is no being that is more encompassing than the multiverse pantheistic God. Therefore, the multiverse pantheistic God is the greatest possible being.

Again, the multiverse pantheistic God is not an agent, so it might not be appropriate to adopt the same religious attitude towards Him as that which traditional theists adopt towards their God. But at least the multiverse pantheistic God satisfies, according to certain interpretations, the two most common criteria for worshipworthiness—that is, being a creator or cause of our universe and being the greatest possible being. Hence, the third objection does not undermine multiverse pantheism.[11]

EVIL AND MULTIVERSE PANTHEISM

As we saw in the previous section, contrary to traditional pantheism, multiverse pantheism has a good explanation of why there is evil in our universe. However, it faces a different form of the problem of evil. If multiverse pantheism is true, God encompasses not only some possible evil states of affairs but *all* possible evil states of affairs, including all possible utterly awful evil states of affairs. This means that multiverse pantheism faces quantitatively and qualitatively more significant evil than traditional theism and traditional pantheism do.

Again, multiverse pantheism adopts the idea that greatness should be understood, not in terms of great-making properties, but in terms of the scope of that which it encompasses. According to this interpretation, the more encompassing a being is, the higher the greatness of that being is. So the amount of evil that God subsumes actually enhances, rather than diminishes, the greatness of God. Conversely, if the multiverse fails to encompass some evil states of affairs (or any state of affairs at all) it fails to qualify as the greatest possible being. Having said that, it is still disturbing to think that God includes utterly awful evil states of affairs, such as innocent people being tortured for a very long time, possibly an infinite amount of time, for no reason. While this might not pose any metaphysical problems for or logical incoherencies in multiverse pantheism, it is a highly disturbing moral implication. In this final section before the conclusion, I consider this problem and possible responses to it. I argue that, unfortunately, none of the responses is satisfactory.

The first response is hinted at by Balbus, Cicero's (1972) character in *The Nature of Gods*, who defends Stoic pantheism:

There is the influence exerted upon us by the great blessing which we enjoy from a temperate climate, from the fruitfulness of the earth, and from the abundance of other blessings . . . There is the awe inspired by thunderbolts, storms, cloudbursts, blizzards, hailstorms, floods, plagues, earthquakes or sudden tremors of the earth, showers of stones, and raindrops red as drops of blood; from the subsidences and sudden fissures in the earth; from monstrosities in man or beast: from fiery portents and comets in the skies, such as recently foretold frightful disasters in the civil war: or from the appearances of twin suns, as happened, so my father told me, in the consulship of Tuditanus and Aquilius, the year in which Scipio Africanus died, himself as glorious as a second sun. Terrified by such events as these, men came to see in them the working of some divine and heavenly power. (129)

Balbus has pantheism in mind in the preceding passage, but one might try to apply his reasoning to multiverse pantheism as well. Perhaps the evil aspect of the multiverse not only fails to undermine multiverse pantheism but also helps to make the multiverse pantheistic God awe-inspiring. Perhaps the multiverse pantheistic God is awe-inspiring because He encompasses all possible forms of evil, including natural disasters. It is, of course, difficult to regard the multiverse pantheistic God as awe-inspiring if He is entirely evil. However, He also encompasses many, in fact infinitely many, good states of affairs. Some of them are the greatest possible states of affairs exhibiting goodness beyond our imagination. Hence, one might claim, the multiverse pantheistic God has two contrasting aspects, good and evil, the combination of which makes Him awe-inspiring.

Unfortunately, this response does not save multiverse pantheism. Traditional pantheism needs to countenance only a finite amount of evil actualized in our universe because the traditional pantheistic God does not encompass evil outside our universe. Balbus's approach, therefore, might be effective for saving traditional pantheism. Perhaps God needs to have a certain finite amount of evil in order to be awe-inspiring. Conversely, perhaps a being that encompasses only good is not awe-inspiring and fails to qualify as God. Yet the same reasoning does not apply to multiverse pantheism because the quality and quantity of evil that the multiverse pantheistic God encompasses constitute far too much evil. He encompasses all possible forms of evil including utterly awful ones such as innocent people being tortured for a long time, possibly an infinite amount of time, for no reason. It is difficult to think that such a being is qualified to be God.

The second response to the problem is parallel to Lewis's defense of modal realism. As I mentioned earlier, the multiverse model on which modal pantheism relies is essentially identical to modal realism. And modal realism is known to face an ethical problem that is relevant to the problem under consideration. The problem for modal realism is that the view appears to discourage us from acting morally because whether or not we act morally and prevent evil in this world, evil is instantiated in some other possible worlds anyway. This is because, given the claim of modal realism that all possible worlds exist, the total sum of good and evil does not change whether or not we act morally in this world. Lewis's response to this problem is that evils in other possible worlds should not bother us in the moral context. He writes,

"For those of us who think of morality in terms of virtue and honor, desert and respect and esteem, loyalties and affections and solidarity, the other-worldly evils should not seem even momentarily relevant to morality. Of course our moral aims are egocentric. And likewise all the more for those who think of morality in terms of rules, rights, and duties; or as obedience to the will of God" (Lewis 1986, 127).[12] One might apply similar reasoning and say that we should not be bothered that there are utterly awful evil states of affairs in universes other than ours. When we talk about what is good and what is evil, one might say, our main concern is morality in *our* universe and our moral concerns should not extend to other universes that are causally and spatiotemporally distinct from ours.

However, if the multiverse model in question is correct, it is difficult not to extend our concerns to other possible universes in our context because even if people in other possible universes are morally irrelevant to what we do in our universe, they nevertheless exist and form part of God as the totality. Adopting an egocentric point of view allows us to set aside other universes *in considering our moral actions* but it does not allow us to set aside other universes *in considering the multiverse identified as God.*

The third response says that the problem in question should not bother multiverse pantheists because, overall, the axiological value of the multiverse is at least neutral, possibly positive, and possibly infinitely positive. The multiverse encompasses utterly evil states of affairs but we should not forget that it also encompasses *all* utterly good states of affairs as well. So, for example, while it includes a state of affairs in which people are tortured for an extended period for no reason, it also includes a state of affairs in which people experience eternal happiness. It is difficult to calculate the overall axiological value of a sum of many universes, but multiverse pantheists can optimistically hope that the overall value is at least neutral, possibly positive and possibly infinitely positive.

But this response is not satisfactory either. Even if the total value of the multiverse is infinitely positive, it is still puzzling why some of the most awful events are part of God. Focusing on the worst possible instance of evil that God encompasses, multiverse pantheism faces the most acute form of the problem of 'horrendous evil'. Marilyn Adams (1999, 26) defines horrendous evil as an instance of evil "the participation in which (that is, the doing or suffering of which) constitutes prima facie reason to doubt whether the participant's life could (given their inclusion in it) be a great good to him/her on the whole". Adams's examples of horrendous evil include "the rape of a woman and axing off of her arms, psycho-physical torture whose ultimate goal is the disintegration of personality, betrayal of one's deepest loyalties, child abuse of the sort described by Ivan Karamazov, child pornography, parental incest, slow death by starvation, [and] the explosion of nuclear bombs over populated areas" (26). The multiverse pantheistic God encompasses all of these instances and He also encompasses worse ones, indeed infinitely worse ones. Even those theists who think that the existence of horrendous evil in our universe is justified in a theistic framework are unlikely to think that the existence of horrendous evil of every possible quality and quantity can be justified.

The fourth, and final, response says that the multiverse does not include all possible universes but all and only possible universes that are overall good. Klaas Kraay (2010) argues that if traditional theism is true, God creates a multiverse comprising all and only those universes that are worthy of creation and sustenance. Multiverse pantheists might apply the same reasoning and hold that God is the multiverse comprising all and only those universes that are worthy of creation and sustenance.

This solution succeeds in eliminating utterly evil states of affairs from God but it also eliminates some distinctive features of multiverse pantheism. First, it makes it impossible for the multiverse pantheistic God to satisfy the Anselmian notion that God is the being than which no greater can be thought, insofar as greatness is construed in terms of encompassment. This also means that we cannot construct the ontological argument for the multiverse pantheistic concept of God, which means that such a God is not a necessary, self-existent being, leaving unanswered the question why such a multiverse exists in the first place. It also implies that the multiverse pantheistic God fails to satisfy the common criteria for being worshipworthy, namely, being the creator of the universe and the greatest possible being. The multiverse pantheistic God in question also cannot be characterized as omniscient, omnipotent and omnibenevolent. We cannot call such a God omniscient, omnipotent and omnibenevolent because this God subsumes only all pieces of knowledge *in some universes,* all instances of power *in some universes,* and all possible forms of benevolence *in some universes.*

Therefore, none of the four responses to the unique problem of evil for multiverse pantheism succeeds in fully disposing of the problem.

CONCLUSION

If there are good scientific and philosophical reasons to think that the world consists of the multiverse, it is natural for pantheists to prefer multiverse pantheism over traditional pantheism. As we have seen, multiverse pantheism appears to be an attractive option initially because it retains some of the features of traditional theism while avoiding several strong objections to traditional pantheism. Multiverse pantheism, at least that version according to which the multiverse consists of all possible universes, faces, however, a disturbing consequence concerning evil which is likely to dissuade many people from endorsing the view. Berkeley says that the problem of evil is everyone's problem and everyone's problem is no one's problem (as quoted in Stoljar 2006, 51). Unfortunately, however, multiverse pantheism seems to face a particularly forceful version of that problem.[13]

NOTES

1. Some explicitly construe pantheism as a thesis about the universe. For example, the *Oxford English Dictionary* defines 'pantheism' as "the religious belief or philosophical theory that God and the *Universe* are identical" (emphasis added).
2. Note that I use the term 'world' rather loosely here. By this term I do not intend to refer to a possible world according to possible world semantics. If God exists, then He must exist in the actual world, that is, one of the possible worlds, so God cannot exist independently of the actual world. What I mean by the term 'world' here, therefore, can be construed as the sum of all existing universes. One might also wonder exactly what the term 'universe' refers to. Does a universe include only physical objects and properties, or does it include nonphysical objects and properties as well? I set this question aside here.
3. Alternatively, one might construe pantheism and pan*en*theism in terms of our universe rather than the world. So according to this construal, pantheism is the view that God is identical with our universe, rather than the world, while pan*en*theism is the view that our universe is only a proper part of God. If we adopt this construal, then multiverse pantheism turns out to be a version of pan*en*theism because it holds that our universe is only a proper part of God and God is identical with the multiverse that includes our universe. God is beyond our universe in the sense that our universe does not exhaust God. However, whether multiverse pantheism should be characterized as a version of pantheism or panentheism is a matter of terminology.
4. As is common practice, I use the pronoun 'He' to refer to God, but this is not ideal for referring to the pantheistic God in particular, as pantheism does not regard God as a person.
5. For discussion of the grounds of God's worshipworthiness, see Bayne and Nagasawa (2006).
6. For a similar point, see Oppy (1997) and Steinhart (2004).
7. The phrase 'at least in one sense' is crucial here. I do not mean, for example, that the universe knows things in the way that we do. Multiverse pantheists would be committing the fallacy of composition if they hold that the facts that I know that p and that the universe encompasses me entails that the universe knows that p in the exact sense in which I know it. Thanks to Klaas Kraay on this point.
8. Notice that, if traditional pantheism is true, God encompasses only all forms of knowledge, power, and benevolence that are instantiated *in this universe*. This hardly entails that God is omniscient, omnipotent, and omnibenevolent.
9. I am deeply indebted to John Leslie on these points concerning the multiverse model and modal realism. See Klaas Kraay (2011) and Nagasawa (forthcoming) for further discussion of possible worlds and the multiverse.
10. It should be noted that Almeida makes this point in relation to the compatibility between traditional theism and modal realism. His focus is not on multiverse pantheism.
11. One might claim that the multiverse pantheistic God is not worshipworthy because we cannot worship a being that is not a person. This raises a further question of exactly what worship is. I do not have space to discuss this controversial issue. My point here is that the multiverse pantheistic God satisfies at least the two most common criteria for being worshipworthy. For further discussion of worship, see Bayne and Nagasawa (2006).
12. For more on the ethical implications of modal realism, see Adams (1979) and Heller (2003).

13. I presented an earlier version of this paper at the God and the Multiverse Workshop (Ryerson University, February 15–16, 2013). I would like to thank Klaas Kraay, who organized the event, and all the participants. I would particularly like to thank Peter Forrest, Klaas Kraay, John Leslie, and Eric Steinhart for their written comments. This essay was written as part of my Templeton project with Andrei Buckareff, titled "Exploring Alternative Concepts of God". I am very grateful to the John Templeton Foundation for its generous support.

REFERENCES

Adams, M. M. 1999. *Horrendous Evils and the Goodness of God*. Ithaca, NY: Cornell University Press.
Adams, R. M. 1979. "Theories of Actuality." In *The Possible and the Actual*, edited by M. Loux, 190–209. Ithaca, NY: Cornell University Press.
Almeida, M. 2011. "Theistic Modal Realism?" *Oxford Studies in Philosophy of Religion* 3: 1–14.
Bayne, T., and Y. Nagasawa. 2006. "The Grounds of Worship." *Religious Studies* 42: 299–313.
Bigelow, J., and R. Pargetter. 1987. "Beyond the Blank Stare." *Theoria* 53: 97–114.
Cicero, M. T. 1972. *The Nature of Gods*. 45 BC. Translated by C. P. McGregor. London: Penguin Books.
Culp, J. 2013. "Panentheism." In *The Stanford Encyclopedia of Philosophy*, edited by E. N. Zalta. http://plato.stanford.edu/entries/panentheism/.
Findlay, J.N. 1955. "Can God's Existence Be Disproved?" In *New Essays in Philosophical Theology*, edited by A. Flew and A. MacIntyre, 47–56. London: SCM Press.
Heller, M. 2003. "The Immorality of Modal Realism, or How I Learned to Stop Worrying and Let the Children Drown." *Philosophical Studies* 14: 1–22.
Kraay, K. J. 2010. "Theism, Possible Worlds, and the Multiverse." *Philosophical Studies* 147: 355–68.
Kraay, K. J. 2011. "Theism and Modal Collapse." *American Philosophical Quarterly* 48: 361–72.
Lewis, D. K. 1986. *On the Plurality of Worlds*. Oxford: Blackwell.
Morris, T. V. 1984. "Duty and Divine Goodness." *American Philosophical Quarterly* 21: 261–68.
Nagasawa, Y. Forthcoming. "Modal Panentheism." In *Alternative Concepts of God*, edited by A. Buckareff and Y. Nagasawa. Oxford: Oxford University Press.
Oakes, R. 1983. "Does Traditional Theism Entail Pantheism?" *American Philosophical Quarterly* 20: 105–12.
Oppy, G. 1997. "Pantheism, Quantification and Mereology." *The Monist* 80: 320–36.
Steinhart, E. 2004. "Pantheism and Current Ontology." *Religious Studies* 40: 63–80.
Stoljar, D. 2006. *Ignorance and Imagination: The Epistemic Origin of the Problem of Consciousness*. Oxford: Oxford University Press.
Swinburne, R. 1981. *Faith and Reason*. Oxford: Clarendon Press.

10 God and Many Universes

John Leslie

MULTIPLE UNIVERSES

The word 'universe' used to mean the sum total of all existing things, with the possible exception of God. Today matters have changed. (a) For example, cosmologists now often classify cosmic oscillations—Big Bang, Big Squeeze, Big Bang, and so on—as giving us a sequence of universes, particularly if the cosmos is pictured as born again with markedly new properties after each Big Squeeze. (b) In the case of Inflation's cosmos, currently very popular, universes with different properties are instead separated not by Time but by Space. The Big Bang speeded up very briefly after its earliest moments, the cosmos doubling in size many times and developing domains with differing characteristics when a unified force broke apart in ways governed by chance. The domains are so huge that everything out to our horizon—it lies as far distant as light could have traveled towards us since the Bang—is deep inside a single domain, 'our universe'. (c) Again, String Theory predicts universes wildly different in their properties—hugely many universes. They may number as many as 10 followed by 499 zeros. (d) Many Worlds Quantum Theory views the cosmos as constantly dividing into 'superposed' branches corresponding to all of the different ways in which quantum events could occur. The branching would have started in the Big Bang's earliest moments, for few cosmologists believe that the existence of observers would be essential to such branching. Each branch could well be called 'a universe'. Once again, the different universes could differ greatly in their properties, thanks to differences dating from the early time when a unified force broke apart. (e) Et cetera. New ways of getting multiple universes are suggested so frequently that it is hard to keep track of them.

All this is far from being pure speculation, totally irrefutable. Physical findings constantly threaten to refute particular ways of getting many and varied universes. For a start, a popular way of developing differences between universes involves *scalar fields* which fall to different minima, the field strengths then dictating the masses of elementary particles (the electron, the proton, and so on) and the strengths of physical forces (gravity, electromagnetism, and so forth). The most popular scalar field for the job is the Higgs field, and for a period the Large Hadron Collider looked very much as if it might rule out its existence: the window in which the Higgs particle could be found was getting ever narrower. Again, there have been vigorous efforts to show that cosmic oscillations are impossible. And although Inflation is very widely accepted, there are interestingly strong reasons for thinking it cannot explain what it was meant to explain, the puzzling cosmic 'flatness' and 'smoothness' (space is nearly or exactly Euclidean, and there is strangely much uniformity in the large-scale distribution of matter plus a remarkable absence of life-excluding turbulence). Roger Penrose has argued that the smoothness is far too great to be explained by Inflation, and that flatness would be vastly more likely *without* Inflation.

Many Worlds Quantum Theory, too, can be all too easily refuted unless developed very carefully. It risks being in disastrous conflict with common sense. True enough, there is little force in the protest that we do not experience the multiplicity of its Worlds. The theory itself appears able to say why anybody only ever experiences a single such World. The risk, though, is that the theory will tell us not to trust such things as calendars and newspapers announcing the day's date. Popular accounts of Many Worlds Quantum Theory typically describe a process of 'splitting' that causes *a continual increase in the number of observers* from each moment to the next. Everybody is described as developing more and more *instances:* trillions upon trillions more with every passing minute. The name 'John Leslie', for example, would apply to more and more separately existing John Leslies, each unconscious of the thoughts and experiences of the others, as the splitting proceeded. But unfortunately this would mean that the actual date was almost certain to be later than what one thought. The fact that so many more instances of oneself would be observing things at later dates would swamp any apparent evidence that the date was really as early as stated by the newspaper or by the calendar on the wall. Resolving this problem requires considerable care. One needs proliferation of observer-*versions* with no increase in number of observer-*instances*. (Rather as in the case of a group of men who become more differentiated through half of them taking off their hats. One hundred men at the start, and one hundred at the finish, but now fifty hatted and fifty hatless.) Many Worlds Quantum Theory as developed by David Deutsch allows for this. Various other versions of the theory are in trouble.

On the whole, however, multiple universe theories have stood up well to attempts to refute them. It is now commonly accepted that there exist immensely many gigantic physical realms, each worth calling 'a universe'. Universes could often be nested inside others—an inflated cosmos, for instance, could be split into enormous regions, 'universes', that were in turn separating into universes of more and more varieties in the way that Many Worlds Quantum Theory describes. There could in addition be universes having absolutely no contact with one another. Several physicists suggest that our universe 'quantum-fluctuated' into existence. Now, a quantum-fluctuational mechanism for creating a universe could scarcely be expected to operate *only once*. If once, then why not up to infinitely many times, producing universes each entirely isolated from the others?

FINE-TUNING

The thing that has done most to make physicists believe in multiple universes is the apparent evidence of Fine-Tuning. Our universe seems 'fine-tuned for life' in the following sense: that very small changes to the masses of its elementary particles, the strengths of its physical forces, its early expansion rate, its degree of turbulence, and so forth, would have made it a universe in which life could never have evolved. All or almost all of what are called 'free parameters' of physics and cosmology—the numbers that are at present entered into the equations 'by hand' instead of being calculated from basic theoretical principles—appear to have needed tuning to some degree, and in several cases very accurate tuning seems to have been required. One case has attracted particular attention. The equations of General Relativity include a number, the cosmological constant, which Einstein set to zero. Quantum theory now suggests that the number would 'naturally' have been vastly greater, 10 to the power of 120 times greater, than is compatible with what we observe. Ten to the power of 120 is a trillion trillion trillion trillion trillion trillion trillion trillion trillion trillion, and the observations which seem so immensely 'unnatural' are all the observations that show that the universe expanded at just the right rate for stars and planets to be able to form. Why was the cosmological constant tiny enough? Theorists once hoped that the quantum effects tending to make it gigantic instead of zero would be exactly counterbalanced by other factors inside some neat theoretical package, but it is now known that the number is in fact *nonzero*, leaving us with the problem that it seems to have needed tuning with an accuracy comparable to throwing a dart to hit a microscopically tiny target placed at the far edge of the observable universe. Just what could account for *that*, let alone for all the other instances of apparent fine-tuning?

Two possible answers suggest themselves. (A) The first involves Divine Selection. Some factor deserving the name 'God', some factor specifically concerned with ensuring that conditions would be suitable for the evolution of intelligent life, selected everything appropriately, either by tuning free parameters very accurately or by very accurately picking some Fundamental Physical Theory leading to a universe which had every appearance of being fine-tuned but whose properties were in fact dictated by that theory so that the parameters were not truly 'free'. (B) The second answer involves a combination of chance variations between universes and Observational Selection. The idea is that universes exist in tremendous numbers and in immense variety. Our universe is one of the perhaps extremely rare ones where conditions allow observers to evolve—but of course if they had failed to allow this then we would not be here to observe it.

Given sufficiently many universes and sufficient variation between universes, could Divine Selection be entirely replaced by Observational Selection? Well, there is the following major difficulty. In order for our universe to be life-permitting, the strength of a physical force (electromagnetism, for instance) or the mass of a particle (the electron, for example) often has to be accurately tuned for several different reasons at once. Divine Selection might well be needed to explain why this leads to no problems.

What problems? Well, force strengths and particle masses which varied randomly between universes could be easy enough to explain, for example by variations in scalar fields. Scalar fields have intensity but no directionality such as characterizes the field which a compass needle can detect, and thanks to Inflation they could have the same intensity right out to our cosmic horizon. We would then be unable to detect them by any straightforward tests, but there are elegant physical theories which explain force strengths and particle masses by postulating such fields, and which suggest that the field intensities would be matters of chance so that they would vary from universe to universe. We would thus be able to understand why in our universe the strength of electromagnetism, for instance, is in the right ratio to the strength of gravity for stars to be able to burn for billions of years in life-supporting ways. It has been argued that in this particular case, the tuning needed to be accurate to one part in many trillions, but given hundreds of trillions of universes and random variations in scalar fields, such tuning might well have been achieved by chance in at least a few universes—and if our universe were not one of them, then we would not be here to observe it. Yet how on earth are we to explain the fact that electromagnetism manages to be tuned in the right way to satisfy *many further requirements as well?* It needed tuning for carbon to be cooked in great quantities in stellar factories without almost all of it then being converted to oxygen; for quarks not to be replaced by leptons, making atoms impossible; for protons not to decay swiftly; and so on down a fairly long list. Why was tuning *to any one strength* able to fulfill *so many different needs?* Why did not electromagnetism have to be tuned in one way to satisfy one requirement, in a very different way to satisfy a second requirement, in yet another way to satisfy a third, and so on? And the same question arises in the cases of other crucial factors. So why, we can well ask, were not *all* possible mixtures of force strengths and particle masses equally life-excluding, with tuning to satisfy any one of life's prerequisites *messing up* tuning to satisfy its other prerequisites?

The difficulty might go away if universes varied greatly not just with respect to force strengths and particle masses, but even in their fundamental laws of physics. Given vastly many universes and great variations in fundamental laws, there might be at least one universe in which this sort of problem did not arise. But while varying scalar fields could be readily explained, variations in fundamental laws cannot so easily be understood—unless one

brings in God. What would otherwise be very odd law-variation from uni-
verse to universe, law-variation very hard to reconcile with the Principle of
Induction which is "Expect More of the Same", could become plausible if
God's will were thought to control what the laws would be.

There could, however, be strong grounds for thinking instead that if God
created more than a single universe, then he chose the same fundamental
laws for all the universes he created, as a straightforward way of ensuring
that all of them would be life-permitting—since wouldn't ensuring it be
what God would have wanted?

HOW MANY UNIVERSES, AND OF WHAT KINDS, COULD BE
EXPECTED IN A GOD-CREATED SITUATION?

Would God create any *lifeless* universes if these existed in isolation instead
of being, for instance, huge lifeless regions inside an immensely inflated cosmos that was life-permitting in at least one region? Some have suggested
that God would indeed create lifeless universes for the fun of seeing them
develop. However, is not God able to know everything knowable, so could
he not know exactly how deterministic universes would develop, viewing
their development "in his mind's eye" without having to create them? And
in the case of *in*deterministic universes, would not God know all the possible ways in which they could develop, once again without having to create
them? But suppose that God wanted to create universes not for the fun of
watching them but for the benefit of the conscious living beings that universes can contain. How many universes would he then create?

The answer might seem to be that God would not limit himself to creating
just one, or just fifty, or just sixty-seven trillion. Wanting to produce as much
good as possible, he would create infinitely many universes. Even if our own
universe extends infinitely far and contains infinitely much worthwhile life,
God would have created infinitely many more such universes *as well* since
each would be a worthwhile addition to the goodness of the situation. To see
this, suppose God had indeed created infinitely many universes, each filled
with infinitely many happy living beings, and then consider how evil it could
be to annihilate all but one universe on the excuse that this last remaining
universe contained infinite goodness, just by itself. Surely it would be nearly
as devilish as annihilating every single universe on the excuse that God's
own existence was enough, just by itself, to make Reality infinitely good. In
Ethics we must not be dazzled by the word 'infinite'. If a thing had infinite
badness, infinite negative value, who would dream of denying that creating
another thing like it would make matters *worse?*

It could further be thought that all the universes which God created would
be life-containing even if many of them were indeterministic. For remember,
it is standard theology to see God as 'conserving' any universe in existence
from moment to moment, almost as if continually creating it anew. Well,

imagine that an indeterministic universe could become life-permitting only if at early moments a unified force split into many forces, all of which were tuned appropriately. Imagine God viewing the situation when the unified force began to break apart. When choosing how to conserve the universe from each moment to the next, always with slight changes, would not God have to choose between the possibilities that the indeterministic laws of the situation left open, and would he not have every reason for choosing in such a way that new forces resulting from the split were indeed tuned appropriately? And similarly later on, if God saw a warm little pond such as Darwin imagined, a pond where a situation had developed in which a few indeterministic events of the right sort, quantum events perhaps, would allow life to originate in the pond. Conserving the universe in slightly changed states from each moment to the next, would not God have reason to pick, from among the undetermined possibilities, the ones which were life's first beginnings? This would scarcely count as the kind of 'interference' that theists think God had good reason to avoid, for instance interference to save the virtuous from having rocks fall on their heads. Of course in a world in which there is always some very faint chance that we shall find ourselves quantum-tunneling through brick walls, God *could* exploit quantum uncertainties so as to save the virtuous from falling rocks. But if he lacked good grounds for avoiding such interference, then there would be no adequate defenses against the theological problem of evil—whereas ensuring that a universe *at least wouldn't be lifeless* is unable to ruin such defenses. The defenses spring into action only once living beings have come to exist.

There does seem, though, to be a gigantic problem which makes the problem of evil utterly beyond solution unless very dramatic changes are made to your typical theist's world-picture. Your typical theist describes God as able to create absolutely anything that is genuinely possible. *And yet* your typical theist, rejecting pantheism, pictures God as creating immensely many things each with an existence fully separate from his own, and each infinitely inferior to himself. Well, why? Why not create other beings like himself, divine beings with mental lives as worth living as his own? Why not create infinitely many of them? Would not that be far better than creating infinitely many universes whose living beings had lives of the dissatisfying sort had by you and me? So when we ask ourselves how many universes a benevolent, omnipotent deity would have created, may not *"Infinitely many universes"* be quite the wrong answer? Should not our answer be *"No universes at all"*?

A PANTHEISTIC PICTURE OF DIVINE MINDS AND OF UNIVERSES

In my book *Infinite Minds* (Leslie 2001a) I defend a cosmos consisting of infinitely many infinite minds, and of nothing else. The thought-patterns of those infinite minds include the patterns of actual universes, there never existing any actual universe anywhere else. Our own universe, for instance, *just is* a pattern contemplated by one of the minds, and when contemplating it the mind in question is not contemplating anything outside itself; the pattern is simply a pattern about which it is thinking in full detail, a pattern carried by that mind because no mind can think of all the details of a complex pattern without itself being complexly patterned. I call all of the minds 'divine'. I picture all of them as having precisely the same characteristics. If, however, you accept Identity of Indiscernibles—the Principle that says that if two things were infinitesimally different, then neither of them could change in a way removing the infinitesimal difference without one of them vanishing or else the two of them fusing into one—then by all means say instead that the minds differ infinitesimally; I'd still call all of them 'divine'. Each mind knows and experiences everything—or if Identity of Indiscernibles is right, then at least *virtually* everything—that is worth knowing and experiencing, there here being no difference between knowing in full detail what it is like to experience something and actually experiencing it.

Not accepting Identity of Indiscernibles, I picture every infinite mind as knowing *everything* worth knowing so that it is absolutely as good as possible—and therefore eternally unchanging since any change would simply make it worse. You and I live in an infinite mind which experiences, among innumerable other things, everything that you and I experience, because otherwise it wouldn't contemplate in full detail the pattern of our universe, yet this fails to imply that the mind in question *would have to change* instead of remaining eternally the same. It fails to imply it for reasons understood by Einstein when he wrote that our world has "a four-dimensional existence" and when he told the relatives of his dead friend Michele Besso that death was not quite the kind of absolute misfortune it is so often thought to be. There is a sense of the word 'time' which allows it to be true that all experiences that ever are had, are had throughout eternal time. An eternally existing Einsteinian four-dimensional block universe is not a contradiction. Beings living inside such a universe could experience change without the four-dimensional block changing, and the block as a whole could continue to exist, eternally unaltered, in a time whose flow consisted simply in the fact that successive alterations to the block—half of it vanishing, for instance, *and then* the other half—were logically possible, even though never occurring.

None of this seems to me in conflict with what common experience tells us about the world. The situation pictured is somewhat like what physicists such as John Barrow have suggested, which is that you and I are patterns inside a gigantic computer used by the members of an advanced civilization for 'simulating' universes. I simply say that we exist inside an infinite divine mind instead of inside a computer. Again, the situation is much like the one pictured in Hindu and Islamic writings which describe us and all the rest of our universe as existing inside the infinitude of Brahman or of Allah. It is also reminiscent of what is pictured by Spinoza, by Hegel and F. H. Bradley, and by those Christian pantheists who think of God's reality as that "in which we live and move and have our being". It is different, however, in that it is a situation containing an infinite number of divine minds, for I argue that this would be far better than there existing *just one* such mind. Sure enough, even a single such mind would contain infinitely much goodness—but if another existed as well, then the infinite goodness of the first would be a very poor excuse for annihilating the second. Remember, in Ethics we must not let ourselves be dazzled by the word 'infinite'.

In the cosmos as I picture it, the sum total of all existing things, there is an infinity of infinities of universes, for each of the infinitely many infinite divine minds carries the patterns of infinitely many universes. The universes are often very different in their fundamental laws. Some could well be entirely lifeless, for even lifeless universes could contain immensely much that was immensely interesting, immensely worth a divine mind's contemplation. But I also see each divine mind as containing large regions devoted to thoughts about things that are not universes. About infinitely many beautiful mathematical truths, for instance. About what we could call 'dream worlds' of great beauty and interest but not constructed in accordance with any laws, let alone laws of physics. About the experiences of humans (and also no doubt of whales, horses, and inhabitants of other galaxies) whose thought-patterns continue to evolve after the deaths of their bodies, beings who come to share more and more of the wonders of divine knowledge. Our having thought-patterns that were among the thought-patterns of an infinite mind would in no way imply that we *didn't really have* minds and thought-patterns worth calling our own, identities worth calling our own identities; you and I would not be illusions; and after our earthly bodies had died, the infinite mind inside which we existed might take joy in letting our minds share more and more of what it knew and experienced.

This world-picture I would reject as utterly fantastic, had I no suitable creation story—a story about what could generate a situation so remarkably good. But I do have what seems to me a plausible story, and if it is essentially correct, then the situation simply has to be as good as the one described. Anything less good would mean the story was wrong. The story can be called either Platonic or Neoplatonic. Variants on it have been influential for almost 2,500 years. It takes its inspiration from a brief passage in Plato's *Republic,* Book Six, where Plato suggests that The Good, itself "beyond being", is responsible for the existence of things. I think it makes sense to suppose that ethical needs, ethical grounds or reasons for the existence of things, *can themselves* be creatively powerful. It might sometimes be useful to do what Nicholas Rescher (2010) does in such books as *Axiogenesis,* calling the creatively powerful grounds or reasons 'axiological' instead of 'ethical', because so many people have the idea that—by sheer definition of the word 'ethical'—ethical needs, ethical grounds or reasons for things to exist, must always be only *moral* grounds or reasons for people to act in particular ways; they even sometimes say that nothing could ever be good or bad, in an ethical sense of 'good' or of 'bad', unless somebody could perform some action that promoted or opposed its existence. But, disliking the word 'axiological' (it is too much "a philosopher's word") and recognizing that Ethics studies more than just the goodness and badness of actions, I think that the grounds responsible for there existing a cosmos, Something rather than Nothing, can best be described as *ethical*. They of course could not be *moral* grounds since those exist only when there already exists a cosmos containing somebody who could act morally or immorally.

Given my world-picture, how should the word 'God' be used? (1) I suggest that 'God' could reasonably mean the entire infinite ocean of infinite minds in which I believe (with, admittedly, only something like 55 percent conviction; I even think it possible that the cosmos simply happens to exist). You might say I was misusing the word 'God'. You could call me an atheist, just as people called Spinoza an atheist when he developed his variety of pantheism. It would not trouble me much. Not yet converted to any religion, I recognize that many who have been converted would prefer to use the word 'God' in other ways and I see little point in fighting them; *words* are seldom worth a fight. I suggest, though, that 'God' can reasonably be used in many different ways. I could happily use this word *instead* as— (2)—a name for the particular infinite mind inside which I picture myself as existing. Or else—(3)—I could use 'God' as a name for the Platonic Principle that the ethical need for the cosmos to exist is eternally responsible for its existence. These distinctions between different uses of the word 'God' could be purely verbal, for people using the word in these various different ways could all agree on the correctness of my worldview.

What is more, yet another way—(4)—of using the word 'God' is suggested by this worldview. The infinite mind which I picture as my country, the country of which I am a part, is a mind that carries my thought-patterns, just as it carries the patterns of such unthinking things as trees and rocks. My thoughts are often very ignorant. When they are, then (which is a point Spinoza recognized) this infinite mind is itself very ignorant in that part of itself that is *me*. But (and this is another point Spinoza appreciated) the infinite mind has a region or an aspect of its being in which it sees all of the infinitely many patterns that it carries "as if in a single glance" and appreciates them with divine personhood. This all-seeing, personality-imbued region or aspect of the infinite mind could particularly well deserve the name 'God', above all because it included love for individual humans and other intelligent living beings.

Here are various further elements of my worldview:

(a) The situation pictured is mental throughout, although large parts of it also count as 'physical' since they are the patterns of physical universes. Being physical is just a matter of having a pattern—a structure—of a particular type, and an infinite mind thinking about a possible physical universe in full detail makes that universe both fully real and fully physical just by thinking about it. (My conviction, remember, is that no mind can think about a pattern in full detail without itself carrying such a pattern, and in my picture the pattern of a physical universe never exists in any 'stuff' that is outside an infinite mind. You can accept all the findings of physics without believing in such 'stuff'.) Now, in *Value and Existence* I defended phenomenalism, the theory that *only some parts* of our universe as commonly described would actually exist, the parts that were thoughts and experiences. This still seems to me a view that cannot be refuted by any experiences or by analysis of how languages are learned. Even in Bishop Berkeley's phenomenalistic universe, languages are learned when God stimulates minds appropriately. But in that first book of mine, in its eleventh chapter, I suggested that there could be infinitely many unified centers of consciousness, each with an existence separate from that of the others and each of unlimited complexity, and it now strikes me that this leads naturally to the world-picture I defend in *Infinite Minds*, of infinitely many minds each of which contemplates (among other things) the entire patterns of infinitely many universes—*not* just the parts of those patterns which phenomenalists would accept.

(b) Would those infinite minds know absolutely all facts, like the God of your typical theist? It seems to me that many facts would not be worth knowing, and therefore would not be known. Even facts about the contents of a single room are related in an infinity of ways—of more and more facts parasitic upon other facts—because all relationships between things give rise to relationships between relationships, and then relationships between relationships between relationships, and so on *ad infinitum*. A divine mind which contemplated all of this would be crammed with endless tedious clutter. It could, for example, be a fact that a teapot in the room was 34.46 times as heavy as one of the teacups in the room, and then a further fact that 34.46 was a number 9.78 times smaller than the length in centimeters of a table in the room, and yet another fact that expressing all this in Spanish words and Arabic numerals would fill at least such and such a number of inches on the screen of the computer in the room when displayed in Times New Roman, font size eleven. An infinite mind, I suggest, would be less than ideally good if it contemplated facts as messy as those. Again, an infinite mind could do well *not* to think about the patterns of all logically possible universes, which would make them into *really existing universes* in my world-picture, just as they are in the modal realist picture developed by David Lewis. It could do well not to think about them because, first, infinitely many of those logically possible universes would contain nothing but beings suffering terrible, infinitely prolonged agonies, and second, the universes that had any marked degree of orderliness would be extremely rare, the others being filled with disorderly rubbish. Even among logically possible universes that had possessed what we call causal orderliness for billions of years, the ones that continued to be orderly for even another millisecond would be rare. This would give us every reason to expect to die immediately if we thought that all logically possible universes really existed—a point that has been made by modal realism's opponents.

(c) I view each infinite mind as fully unified in its existence. Its parts would be no more capable of existing separately than are the color, the length and the hardness of a stone, all of which are abstractions from the stone's concrete reality in which these items are fused together. Now, you might perhaps ask, would it not be *better* if my infinitely many separately existing infinite minds, minds forming the infinite ocean that I am willing to call 'God', were replaced by a God of a more traditional kind, a single divine being, through the fusing together of those minds? Well, my answer is a firm No, because Identity of Indiscernibles does at least apply to the qualities of any single thing. A single billiard ball, for example, cannot be green all over *and also* green all over. It can be green all over only once. Now, if those infinitely many minds were all of them identical, then fusing them would result in a situation exactly the same as if all but one of them had been annihilated, for to imagine them as remaining distinct from one another despite the fusion would be like imagining a billiard ball which was green all over *infinitely many times*. Yet suppose, instead, that Identity of Indiscernibles had forced the minds to be different, so that only one of them knew and experienced *absolutely everything* worth knowing and experiencing, while each of the others experienced all of this with the exception of some infinitesimal ingredient—a different infinitesimal ingredient in each case so that Identity of Indiscernibles was not violated. Fusing the minds would then result in a situation exactly as if all of them had been annihilated apart from the very best of them, the one which knew and experienced *everything* that was worth knowing and experiencing. Well, this would be a disaster. Infinitely much that was worthwhile would have been annihilated.

Does this make me a polytheist, a believer in many gods? Certainly, if 'a god' means *a mind worth calling 'divine', existing somewhere*. But *not* a polytheist who sees gods all over the place: one in every mighty river, perhaps, or one in every star. And if the word 'God' stands for my infinite ocean of infinitely many separately existing infinite minds, then I believe in one God only.

MORE DETAILS OF THE PLATONIC APPROACH TO
WHY ANY DIVINE MIND WOULD EXIST

It is odd that a Platonic account of God's necessary existence comes as such a shock to so many of today's philosophers. Even when believing in God and describing God's existence as absolutely necessary, they reject what seems to me a thoroughly traditional way of throwing light on the necessity. One throws light on it *by tying it to God's goodness,* theorizing that the goodness of God is an eternal ground or reason for God to exist, a ground or reason that actually explains why God exists. The philosophers protest that the concept of an ethical ground or reason for the existence of a thing *is different from* the concept of a ground or reason which explains the actual existence of that thing. Well, that's of course right, but why do they consider it a valid basis for a protest? Have they never come across the idea of *synthetic necessity?* Necessity, that is to say, which is just as firm as logical, conceptual necessity, but which cannot be proved by examining concepts.

There is nothing insulting to God in the idea that some things would be outside God's power. Ensuring that two and two added up to five, for example. Or doing something that *made* two and two add up to *four.* Or making it the case that God's existence really was possible; for how on earth could *that* depend on God's being there to will it, to 'ground' it, to be its 'truthmaker'? Or making it true that God's existence was unbeatably wonderful, more ethically needful than the existence of anything else, and that this was the reason why God exists, a reason that was creatively powerful—able, in other words, to account for God's eternal existence— through synthetic necessity. Notice that saying that an unbeatably strong ethical requirement, not itself a product of God's power, was responsible for God's existence, would be *only verbally different* from saying that God's existence was due to one of God's very own characteristics, God's eternal ethical requiredness. Just any theist who writes that God's existence "is due to a necessity of God's own nature" can be described as holding that there is *a requirement* that God's nature be the nature not just of a mere possibility, but of something existing in the way that you and I exist. Yet this would not say that something 'outside God', 'superior to God', was responsible for God's existence. The requirement would have its basis in God's own nature which made it required that God exist. And all this could be true whether God was just a single infinite mind or instead an ocean of infinitely many infinite minds, and whether or not the requirement had an ethical side to it.

Here are five main elements of my Platonic creation story:

(i) The concept of a thing's having intrinsic value is the concept of there being an ethical or 'axiological' need, weak or strong, for the thing to exist, because of the thing's very own nature. (Any such thing's existence, all by itself, would always at least be better than there existing nothing. The thing could have unacceptable consequences, however, or the existence of something else instead could be better.)

(ii) It isn't logically, conceptually necessary that an ethical/axiological need, even for the existence of an unbeatably good situation, is ever by itself sufficient to explain that situation's existence.

(iii) It is nevertheless logically possible for the need for some unbeatably good situation—perhaps there existing infinitely many infinite minds, ones knowing and experiencing everything worth knowing and experiencing—to be sufficient. Its sufficiency would involve no logical, conceptual contradiction. And *either* its sufficiency would be necessary (albeit not logically necessary) *or else* it would necessarily *not* be sufficient; for it would be absurd to see the sufficiency or the lack of sufficiency as just a matter of chance. One way or the other, there would be a necessity here, a synthetic necessity. The case is like that of intrinsic value conceived in the traditional way and without any consideration of possible creative power. Is such value ever real, or is it always a fiction? Logic cannot answer this question, but a necessity is involved nevertheless. *Either* at least one thing has some degree of intrinsic value as traditionally conceived *or else* such value is a fiction; and whichever of these alternatives is the case, is the case necessarily rather than through chance, since it would be absurd to fancy that in one world a thing had value of this type while in another world an exactly similar thing had none or was intrinsically bad.

(iv) Platonists need not search for any mechanism that *makes* ethical/axiological grounds for the existence of various things, ethical/axiological requirements that they exist, into creatively powerful realities—realities that themselves account for the coming into existence of those things, or for their eternal existence. Either those requirements necessarily are creatively powerful, or else they necessarily lack creative power, and these alternatives *are equally simple* because neither of them could involve cogwheels whirring, magic wands twirling, angels exerting willpower, or any such other 'explanatory factors'. Cogwheels, wands, and angels are not needed to get synthetic necessities to operate. They operate because they are necessities.

(v) Because only a synthetic necessity would be involved, there is a perfectly acceptable sense of the words 'as such' in which no ethical/axiological requirement *as such* could ever act creatively. It is the sense in which no cow as such is brown. Yet in another sense cows as such often are brown, instead of being white cows coated with brown paint.

REACTING TO THE PROBLEM OF EVIL

Here are four ways in which my world-picture can be defended against 'problem of evil' attacks:

1. Almost certainly, ours would not be the very best of the infinitely many universes in the divine mind inside which it existed. Efforts to portray it as having bestness are therefore misguided. It could even be wrong to call it better than any 'dream world', a world not governed by physical law—for infinitely many such worlds might exist inside each divine mind, and our universe might be inferior to many of them. All that we need suppose is that universes like ours do add something worthwhile to the divine minds inside which they exist.

2. You and I have plenty of room for *moral efforts,* struggles to make our universe better than it would be if we failed to struggle, whether or not its events are fully deterministic (I here accept standard 'compatibilist' arguments whose main point can be stated without even mentioning the word *freedom*). Similarly, there is room for such struggles, whether or not there is a sense in which all events, past, present and future, have an existence that is eternal. Einstein thought there was such a sense, yet this did not destroy his grounds for running to try to catch trains.

3. Part of the reason why we would have room for moral efforts is that *not all* goods can be had simultaneously. Recognizing this is common to all reasonable attempts to counter the problem of evil. It is said, for instance, that miracles to prevent wicked actions or disastrous accidents, or to feed the hungry, would erode the good of freedom, and that this could well have *overruled* a benevolent Creator's reasons for producing such miracles; the one ethical need, for the good of freedom, could well have been stronger than the other. Now, many theists who make use of this idea then do something that is wildly inconsistent. Turning round and attacking the Platonic theory that ethical needs could themselves be creatively powerful, without help from any benevolent Creator, they protest that ethical needs *are often not satisfied* so that there are wicked actions, there are disastrous accidents, and there are hungry people. But if wicked actions, disastrous accidents, hungry people, cannot disprove the existence of a benevolent Creator, then how on earth could they disprove the theory that ethical needs themselves created the cosmos?

4. Suppose somebody protests that a minuscule change in the Big Bang would have prevented the Lisbon Earthquake. Theorizing that *absolutely all* worthwhile universes exist inside a divine mind, I can reply that the universe featuring the minuscule change would be a universe different from ours. It could very well exist *in addition to* ours. Well, would it then be good for our universe to be annihilated, instead of continuing to exist side by side with the one in which that earthquake was prevented? Surely not.[1]

NOTE

1. For further development of these ideas, please see Leslie (1978, 1979, 1988, 1989, 1993, 1995, 1996a, 1996b, 1997, 2000a, 2000b, 2001a, 2001b, 2003,

2007a, 2007b, 2009, 2012, forthcoming). Similar ideas have been developed recently by others. For a defense of the "monist" theory that a single whole, mental throughout, carries all the vastly complex pattern of our universe, see Sprigge (1983). For the view that God's existence is due to the ethical need for God to exist, see Ewing (1973), Polkinghorne (1994), Forrest (1996), and Ward (1996). For the idea that God might instead be viewed as the Platonic Principle that ethical grounds for the cosmos to exist are responsible for its existence without the aid of a divine person, consult Forrest (1996), Wynn (1999), Rice (2000), and Rescher (2000, 2010), although Rescher prefers to call the grounds 'axiological' instead of 'ethical'. For there being at least no contradiction in the idea that ethical needs could themselves be creatively powerful, see Chapter 13 of Mackie (1982) and Appendix D of Parfit (2011).

REFERENCES

Ewing, A. C. 1973. *Value and Reality.* London: George Allen and Unwin.
Forrest, P. 1996. *God without the Supernatural.* Ithaca, NY: Cornell University Press.
Leslie, J. 1978. "Efforts to Explain All Existence." *Mind* 87: 181–94.
Leslie, J. 1979. *Value and Existence.* Oxford: Blackwell.
Leslie, J. 1988. "How to Draw Conclusions from a Fine-Tuned Universe." In *Physics, Philosophy, and Theology,* edited by R. J. Russell, W. R. Stoeger, and G. V. Coyne, 297–311. Vatican City: Vatican Observatory.
Leslie, J. 1989. *Universes.* London: Routledge.
Leslie, J. 1993. "A Spinozistic Vision of God." *Religious Studies* 29: 277–86.
Leslie, J. 1995. "Cosmology," "Cosmos," "Finite/Infinite," "World," "Why There Is Something." All entries in *A Companion to Metaphysics,* edited by J. Kim and E. Sosa. Oxford: Blackwell.
Leslie, J. 1996a. "A Difficulty for Everett's Many Worlds Theory." *International Studies in the Philosophy of Science* 10: 239–46.
Leslie, J. 1996b. *The End of the World: The Science and Ethics of Human Extinction.* London: Routledge.
Leslie, J. 1997. "The Anthropic Principle Today." In *Final Causality in Nature,* edited by R. Hassing, 163–87. Washington, DC: Catholic University of America Press.
Leslie, J. 2000a. "Our Place in the Cosmos." *Philosophy* 75: 5–24.
Leslie, J. 2000b. "The Divine Mind." In *Philosophy, the Good, the True, and the Beautiful,* edited by A. O'Hear, 73–89. Cambridge: Cambridge University Press.
Leslie, J. 2001a. *Infinite Minds.* Oxford: Oxford University Press.
Leslie, J. 2001b. "The Meaning of 'Design'." In *Cosmic Questions,* edited by J. B. Miller, 123–38. New York: New York Academy of Sciences.
Leslie, J. 2003. "Cosmology and Theology." In *The Stanford Encyclopedia of Philosophy,* edited by E. Zalta. http://plato.stanford.edu/archives/sum2003/entries/cosmology-theology.
Leslie, J. 2007a. "How Many Divine Minds?" In *Consciousness, Reality and Value: Essays in Honour of T. L. S. Sprigge,* edited by P. Basile and L. B. McHenry, 123–34. Frankfurt: Ontos Verlag.
Leslie, J. 2007b. *Immortality Defended.* Oxford: Blackwell.
Leslie, J. 2009. "A Cosmos Existing through Ethical Necessity." *Philo* 12: 172–87.
Leslie, J. 2012. "A Proof of God's Reality." In *Gottesbeweise als Herausforderung für die Moderne Vernunft,* edited by A. Hutter, F. Hermanni, T. Buchheim, and C. Schwöbel, 411–27. Tübingen: Mohr Siebeck.

Leslie, J. Forthcoming. "A Way of Picturing God." In *Alternative Concepts of God,* edited by Y. Nagasawa and A. Buckareff. Oxford: Oxford University Press.

Mackie, J. L. 1982. *The Miracle of Theism.* Oxford: Oxford University Press.

Parfit, D. 2011. *On What Matters.* Oxford: Oxford University Press.

Polkinghorne, J. 1994. *The Faith of a Physicist.* Princeton, NJ: Princeton University Press.

Rescher, N. 2000. *Nature and Understanding.* Oxford: Oxford University Press.

Rescher, N. 2010. *Axiogenesis.* Lanham, MD: Lexington Books.

Rice, H. 2000. *God and Goodness.* Oxford: Oxford University Press.

Sprigge, T. L. S. 1983. *The Vindication of Absolute Idealism.* Edinburgh: Edinburgh University Press.

Ward, K. 1996. *Religion and Creation.* Oxford: Oxford University Press.

Wynn, M. 1999. *God and Goodness.* London: Routledge.

Multiverses and
the Incarnation

11 Extraterrestrial Intelligence and the Incarnation

Robin Collins

> God planned to create many distinct things, in order to share with them and reproduce in them his goodness. Because no one creature could do this, he produced many diverse creatures, so that what was lacking in one could be made up by another; for the goodness which God has whole and together, creatures share in many different ways. And the whole universe shares and expresses that goodness better than any individual creature.
>
> —Thomas Aquinas, *Summa Theolgiae*

This essay addresses the compatibility of the Christian doctrine of the Incarnation with the hypothesis that there are many other races of vulnerable, embodied conscious agents (VECAs) that are causally isolated from humans and each other. Specifically, as I will define them, VECAs are conscious, embodied beings that are subject to some kind of physical laws; can significantly interact with each other; often choose based on what they think are moral criteria; and who are highly vulnerable to their environment, to each other, and to moral corruption. Further, by being causally isolated from our reality, I merely mean that there is no causal interaction between the races of VECAs in a premortem state. Given these definitions, I will first present reasons for thinking that there are an enormous number of races of VECAs that are causally isolated from each other, and that if God the Son incarnated in our world, then we should believe that he also incarnated in many of these races. I will then argue that, except for the kenotic model, all the major current models of the incarnation in contemporary philosophical theology that claim to be orthodox can easily accommodate the Son's incarnating in many of these other races; and, I will argue, even the kenotic model might be able to accommodate the existence of such races.

There are various types of spatiotemporal regions in which these other races could exist. As three example scenarios, the universe could be very large or infinite and thus contain a very large, if not infinite, number of other planets with VECAs; or, there could be many distinct regions of space-time (a so-called multiverse) generated by some physical process, such as the much discussed inflationary-superstring scenario, with VECAs existing in those other regions;[1] or there could be other realities radically different from ours, as for example in C. S. Lewis's *The Chronicles of Narnia*.

Before proceeding, a few words should be said to motivate the idea that there are other races of VECAs. First, there are motivations that arise out of physics, such as the aforementioned scenarios: if there are enough other planets in our universe, or a multiverse, it is almost certain that VECAs would evolve, unless their existence would require some special divine action—for example, to bring about an immaterial soul connected to a body, a view of the origin of the human soul that Christians have traditionally held. In the latter case, however, it would be odd for God to create a reality with an enormous number of planets that are suitable for the evolution of VECAs, but with God only deciding to intervene so VECAs could exist on one of them.

I do not consider the physics-based motivations at present compelling, however, since claims about how much larger the universe is than our visible universe and whether a multiverse exists are still highly controversial in the cosmological community. Further, although we know that the visible universe contains around three hundred billion galaxies and around a hundred billion stars per galaxy—and hence probably an enormous number of habitable planets—we do not have any good idea of how probable it is for life to develop on a habitable planet by purely naturalistic means, with some, such as Meyer (2010), arguing that it is enormously improbable for the first cell to form even under ideal conditions.

Second, there are theological reasons for believing in an enormous number of races of VECAs. I will sketch one such line of reasoning based on God's perfect goodness; the sketch is not intended to provide a proof of the conclusion, but only to give the reader the basic idea of one important type of theological motivation for it. To begin, we should expect a perfectly good being to create a reality that positively, if not optimally, realizes value—particularly moral and aesthetic value. Given this, presumably, if God planned that human beings exist, then it was because overall our existence positively contributes to the value of reality. If the value realized is primarily intrinsic to what we are and are destined to become, then everything else being equal, a race of VECAs causally isolated from ours but in relatively similar circumstances would also have intrinsic value. If we never interact with each other, than a reality containing us plus one such additional race of VECAs would realize more intrinsic value than a reality without that additional race. The same argument applies to adding a third such causally isolated race of VECAs, and so forth. If the total value contributed by these causally isolated races of VECAs is merely the sum of the intrinsic value of each race, it seems clear that if there is a best of all possible realities, it will contain an infinite number of such races.

The preceding argument for there being an enormous number of races of VECAs is by no means completely compelling, since among other reasons it is possible that there is a further extrinsic value or disvalue that is over and above the sum of the intrinsic value of each race. For example, perhaps God has an aesthetic enjoyment in contemplating these races that depends on there not being too many of them. Or perhaps the value realized by humans and other races of VECAs is partly constituted by gaining more and more knowledge in the afterlife of the nature of reality. If so, it is possible that a race of VECAs learning that there are more than some number N of other races of VECAs would begin to have a negative value for that race by undercutting their sense of being special. These possibilities suggest that the value of reality might become maximal at some finite number of races of VECAs. At least for me, it is hard to believe that it would become maximal with just human beings. If not, then any finite number seems as plausible as any other for when maximality would occur. If, as I have advocated elsewhere (Collins 2005), one should not require that epistemic probability obey countable additivity, then for any finite integer k, one should assign zero epistemic probability of the number of races of VECAs being k.[2] It then follows that for any finite number N, there is zero epistemic probability of the number of races being less than N, and an epistemic probability of one for the number being greater than N. If so, then one should take the position that if there are only a finite number of races of VECAs, then for any finite number N, we should be certain that it is greater than N. Put another way, if we had to bet on whether the number of races of VECAs was less than N or greater than N, we would always give zero odds to its being less than N. So, for whatever number N that one chooses—say a trillion, trillion, trillion—one should bet that it is greater than that, at least if one shares the same sort of intuitions that I have.

If the preceding reasoning is correct, we should think that it is very likely that there is an enormous number of other races of VECAs. In that case, it becomes important for orthodox Christians to examine whether there are views of the Incarnation that are compatible with the statements about the Incarnation made by the New Testament and the ecumenical councils and that allow for such a multitude of races of VECAs. Further, such an examination would give us good reasons to reject views that did not allow for such a multitude.

INCARNATION AND OTHER VECAS

The main concern the existence of other races of VECAs poses for the doctrine of the Incarnation centers around whether this doctrine can accommodate God the Son becoming incarnate in many other races. The issue arises because it seems highly implausible to think that if there are many other races of VECAs, we are the only ones in which God the Son became incarnate. A brief argument for this claim goes as follows. It is obvious that human beings are vulnerable to their environment, to each other, and to moral evil and corruption. Further, presumably God determined that it was a good thing for VECAs like us to exist, and not to intervene to prevent them from becoming "fallen".[3] If this is right, then the preceding arguments based on God's goodness apply to God's creating a reality with many races of fallen VECAs. Thus, if we think it is likely that there are an enormous number of other races of VECAs, we should think it is likely that there is a very large number of races of *fallen* VECAs. Further, under the kind of scenarios arising from physics, the other VECAs arose through an evolutionary process in a physical world. We should, therefore, expect their wills to be vulnerable to the desires arising from their evolutionary origins—such as self-survival, sexual desires, and the like—and hence easily fall into moral wrongdoing.

Now consider two hypotheses, H1 and H2. H1 is the hypothesis that God the Son becomes incarnate in most races of fallen VECAs, whereas H2 is the hypothesis that the incarnation happens in only one such race. Further, let k* contain the information that there are at least N other races of fallen VECAs, and everything else relevant about our race except any information—other than the fact that we are fallen VECAs—that gives us reason to believe the incarnation happened in our world; so, for instance, k* does not contain information such as that there exists a book called "the Bible" which claims that God the Son became incarnate. Finally, let E be the information that God the Son became incarnate in our world. By the odds form of Bayes's theorem of the probability calculus,

$$\frac{P(H1 \mid E \,\&\, k^*)}{P(H2 \mid E \,\&\, k^*)} = \frac{P(H1 \mid k^*)}{P(H2 \mid k^*)} \times \frac{P(E \mid k^* \,\&\, H1)}{P(E \mid k^* \,\&\, H2)},$$

where P(A|B) represents the conditional epistemic probability of a proposition A on a proposition B—which for our purposes can be thought of as the amount a rational human being should believe in A given that she/he believed in B. P(H1|k*) and P(H2|k*) are the *priori* probabilities of H1 and H2, respectively—that is, the epistemic probabilities just on k*, prior to learning information E. On the other hand, P(H1|k* & E) and P(H2|k* & E) are the *posterior* probabilities of H1 and H2, respectively, since they are the epistemic probabilities after one comes to know the evidence E. Finally, $\frac{P(E \mid k^* \,\&\, H1)}{P(E \mid k^* \,\&\, H2)}$ is the Bayes's factor; it determines the factor by which the ratio of posterior probabilities is greater or less than the ratio of prior probabilities. The Bayes's factor, therefore, determines the degree to which the E confirms or disconfirms H1 over H2: if it is greater than one, the evidence confirms H1 over H2, with the confirmation increasing as the Bayes's factor increases; conversely, if it is less than one, it disconfirms H1 over H2.

Now, by stipulation, k* does not contain any information that gives us reason to think that God the Son would be incarnated in the human race instead of any other races of fallen VECAs. Thus, if we knew there were N races of fallen VECAs, under H2 the least arbitrary epistemic probability we could assign to God the Son becoming incarnate is an equiprobability distribution over the races—that is, the least arbitrary probability would give God's becoming incarnate in some particular race, R, of VECAs the same probability as that for any other race R*. Consequently, if we believe that the incarnation only occurs in one race, then conditioned on k*, we should assign the probability of that race being humans as 1/N: that is, P(E|H2 & k*) = 1/N. Since as N gets larger, P(E|H2 & k*) decreases, if we knew that there were at least N other races of VECAs, P(E|H2 & k*) ≤ 1/N. On the other hand, if we believe that the Son becomes incarnate in most races, then we should think that the probability of the Son becoming incarnate in the human race is greater than 50 percent: that is, P(E|H1) > 0.5. Consequently,

$$\frac{P(H1 \mid E \& k^*)}{P(H2 \mid E \& k^*)} = \frac{P(H1 \mid k^*)}{P(H2 \mid k^*)} \times N / 2.$$

The preceding equation entails that for very large values of N, H1 will be very strongly confirmed over H2; so strongly confirmed that even if we give a very low prior probability to the hypothesis that the Son has incarnated in most fallen races, the posterior probability of H1 will be much larger than that of H2: that is, $\frac{P(H1 \mid E \& k^*)}{P(H2 \mid E \& k^*)}$ >> 1, where ">>" means much, much greater than. For example, suppose our prior probability of H1 being true on k* is 0.1 percent and that of H2 being true is 99.9 percent: that is, P(H1|k*) = 0.001 and P(H2|k*) = 0.999. If our background information is that there are at least two billion other fallen races of VECAs, N/2 will be one billion. That will make $\frac{P(H1 \mid E \& k^*)}{P(H2 \mid E \& k^*)}$ one million, which means H1 would be one million times more probable than H2 given the evidence E that God the Son became incarnated in our world. Consequently, if we believed that either H1 or H2 was true, the posterior probability for the human race being the only one that the Son became incarnate in would be about one in a million.[4]

The preceding argument combined with the argument offered previously for an enormous number of fallen VECAs gives Christians substantial reasons for thinking that there is a vast number of other incarnations. Next I will examine the current major views of the Incarnation, and consider which ones can accommodate many incarnations.

THE DOCTRINE OF THE INCARNATION

The central tenets of the orthodox formulation of the Incarnation are given by the ecumenical Council of Chalcedon in AD 451. According to this council:

So, *following the saintly fathers,* we all with one voice teach the confession of one and the same Son, our Lord Jesus Christ: the same perfect in divinity and perfect in humanity, the same truly God and truly man, of a rational soul and a body; consubstantial with the Father as regards his divinity, and the same consubstantial with us as regards his humanity; like us in all respects except for sin; begotten before the ages from the Father as regards his divinity, and in the last days the same for us and for our salvation from Mary, the virgin God-bearer as regards his humanity; one and the same Christ, Son, Lord, only-begotten, acknowledged in two natures which undergo no confusion, no change, no division, no separation; at no point was the difference between the natures taken away through the union, but rather the property of both natures is preserved and comes together into a single person and a single subsistent being; he is not parted or divided into two persons, but is one and the same only-begotten Son, God, Word, Lord Jesus Christ, just as the prophets taught from the beginning about him, and as the Lord Jesus Christ himself instructed us, and as the creed of the fathers handed it down to us. (Tanner 1990, 86)

In brief, the Chalcedon formulation states that Jesus Christ is fully divine and fully human, being one person with two natures. It further asserts that these two natures are fully united with each other—one nature being fully divine and one nature being fully human. Later ecumenical councils refined and clarified these statements, culminating in the Third Ecumenical Council of Constantinople (AD 681). Specifically, this latter council asserted that Jesus also had two "natural wills" and two natural sources of action, with the human will being subject to the divine will; moreover, the council stated that Jesus had a human soul (Ward 1994, 261). The Third Council of Constantinople was the last ecumenical council dealing with the doctrine of the Incarnation, and thus served as the final ecumenical pronouncements on what constitutes orthodox belief regarding this doctrine. Eastern Orthodox and Roman Catholics fully accept the authority of these councils, while most Protestant denominations (that claim to be orthodox) either accept their authority or hold their statements in high regard.

The post-Constantinople views of the incarnation fall into two broad categories regarding the entity which acts, and is the subject of, Jesus' actions, corresponding to what Karl Rahner has called *descending Christology* and *ascending Christology*.[5] In descending Christology, the agent of Jesus' actions and the entity that has Jesus's limited, human beliefs is God the Son. Thus, for instance, when Jesus acts, it is really God the Son that is forming the intention to act and then acting. Since the Council of Chalcedon asserted that there is only one person in Jesus, to explain how Jesus could also be fully human if all of his actions are those of God the Son, advocates of this view claim that the Son assumed or united himself with a human nature. Under this view, some claim that Jesus assumed a human nature while retaining the divine attributes, while others deny this. Among the former fall the "two-minds" and related views discussed later, and the view that Timothy O'Connor and Philip Woodward put forward in Chapter 12 of this volume. Among the latter are the various views falling under so-called kenotic Christology. Each of these will be discussed in more detail in what follows.

In ascending Christology, the entity that is the direct source of Jesus's actions and who has Jesus's limited, human beliefs is a human being. I will consider two versions of ascending Christology later, what Keith Ward has labeled the *enhypostatic* model and the related view of Karl Rahner. For Ward and Rahner, Jesus is divine in the sense that he was united to, or mediated, the Son of God in as complete and perfect a way as is humanly possible in this life; further, Jesus so fully mediates God the Son that when Jesus acts, God the Son could also be said to act, and hence it is appropriate to say that Jesus is God the Son and that God the Son became incarnate in Jesus

My overall conclusion will be that only the kenotic view presents any obvious potential philosophical or theological problems for the Son becoming incarnate in other races. I will start by looking at the two-minds and related views.

The Two-Minds View

As noted earlier, one major way the Chalcedonian statements about the incarnation have been interpreted is in terms of what Thomas Morris calls the *two-minds view* (Morris 1986, 102–7; 1989, 110–27), a view also developed and advocated by Richard Swinburne (1994, 192–215) and philosopher-theologian David Brown (1985, 260–67), among others. For these authors, the "two natures" of Chalcedon and the "two wills" of Constantinople are two systems of consciousness, beliefs, and intentions. Specifically, God the Son takes on a human mind while retaining a completely divine mind that has all the attributes of omniscience, omnipotence, omnipresence, and the like. Says Morris (1989, 121),

In the case of God Incarnate we must recognize something like two distinct minds or systems of mentality. There is first what we can call the eternal mind of God the Son, with its distinctively divine consciousness, whatever that might be like, encompassing the full scope of omniscience, empowered by the resources of omnipotence, and present in power and knowledge throughout the entirety of creation. And in addition to this divine mind, there is the distinctly earthly mind with its consciousness that came into existence and developed with the human birth and growth of Christ's earthly form of existence. The human mind drew the visual imagery from what the eyes of Jesus saw, and its concepts from the languages he learned.

Similarly, according to Swinburne (1994, 202),

> A divine individual could not give up his knowledge, and so his beliefs; but he could, in becoming incarnate in Christ and acquiring a human belief-acquisition system, through his choice, keep the inclinations to belief resulting therefrom to some extent separate from his divine knowledge system. The actions done through the human body, the thoughts consciously entertained connected with the human brain . . . would all be done in the light of the human belief system.

Swinburne calls his view a "divided-mind view" (208), acknowledging that it is similar to the views developed by Morris (1986) and Brown (1985). (See Swinburne 1994, 202n11).

Both Morris and Swinburne present analogies for how this could occur, though they recognize that their analogies are imperfect. Morris (1989, 122–23), for example, offers the analogy of a person voluntarily bringing about a "split personality" and analogies in everyday life in which we are able to divide our consciousness among two tasks. As Swinburne makes clear in his account, God the Son is always the subject (that is, "the center of consciousness") experiencing these two systems of consciousness, and God the Son is always the agent acting—such as moving Jesus's mouth, arms, and legs. Further, insofar as the divine beliefs conflict with the human beliefs, the divine beliefs are the genuine beliefs of God the Son. The human system of beliefs, intentions, desires, and experiences are only taken on in the sense that they are what guide God the Son's actions with regard to the human body and mind of Jesus, with the restriction that they never lead to moral wrongdoing. Although neither Morris nor Swinburne offer the following analogy (perhaps because it brings out a seeming inadequacy of their view), their basic view could be seen as analogous to what happens in the case of *method acting:* one attempts to take on the character's system of mentality—beliefs, intentions, experiences, and the like—and then let it guide one's actions *on stage*. The more successfully one can do this, the more convincing the performance. Given that one does not get lost in the character, one could fully retain one's own beliefs regarding who one really is while playing a character. In an analogous way, God the Son takes on the 'character' of the human Jesus, while still retaining complete consciousness of himself as God the Son and performing the activities of God the Son, such as upholding the entire universe.[6]

As Morris (1986, 163–86) argues in an extensive discussion of the possibility of incarnations in extraterrestrial races, there is no problem under this view of the existence of other races of fallen VECAs. Given the infinite nature of God's overarching divine consciousness, God could take on the nature of all races of fallen VECAs without diminishing the divine consciousness. Indeed, since the infinity of God is typically claimed to 'infinitely' transcend any mathematical infinity, God's overall consciousness would not in any way be diminished even if God took on an infinite number of finite mental systems, from an infinite number of fallen races; in fact, if anything, it would enhance God's consciousness. Thus, where necessary, the Son could accomplish his atoning work in relevantly the same way as he does for humans.

The Enhypostatic View

Keith Ward, former Regius Professor of Divinity at the University of Oxford, rejects the two-minds view in favor of another view in which Jesus is an individual subject distinct from God the Son, but is as completely as possible interpenetrated with the divine nature from the beginning of his existence. Thus, just as in standard Eastern Orthodox theology, humans are called to be interpenetrated/united with the divine life as fully as humanly possible, the human Jesus is as fully interpenetrated with divine life as possible from birth, though how this unity with God played out in Jesus's life was subject to Jesus's own choices and the human developmental stages he went through. Ward calls this the *enhypostatic* view, following a term introduced Leontius of Byzantium in the sixth century (Ward 1994, 272).

Ward (1994, 265–73) presents essentially four motivations for the this view. First, he argues that even though some of the statements made about Jesus in the various councils seem to be claiming that the agent doing the acting is that of God the Son, this could not be a correct interpretation because if it were God the Son, then God the Son would have to be able to undergo change and be affected by the world; it was held by virtually everyone during the period of the ecumenical councils, however, that God was timeless and impassible, and hence could not undergo change or experience suffering.[7] Second, Ward points out that the Third Council of Constantinople declared that Jesus had a human soul; Ward then claims that a human system of mentality (without a human subject), as in the two-minds view, does not seem sufficient to constitute a human soul.[8] Third, and related to the last reason, the two-minds view makes it seem as though God the Son, and hence Jesus, was not really human, but merely play-acting the part, as suggested by the thespian analogy in the last section. Finally, Ward claims that his view allows Jesus to be the "future fulfillment of all of humanity in the Divine Life" (267), and in some real sense the "firstborn" and "first fruits" of transformed humanity, as stated in Romans 8:29 and 1 Corinthians 15:23. Arguably, this role of Jesus does not fit as well with the other views, since the subject of Jesus' actions is not a human subject like us.

One worry with Ward's view is that it undercuts, or at least deemphasizes, the sense in which Jesus was really the Son of God incarnate, as many take to be implied by New Testament passages such as John 1:1,14 and Philippians 2:6–8.[9] Ward, however, attempts to account for passages like these by arguing that the interpenetration of Jesus by the divine Logos was so extensive and deep that it was proper to treat the actions and life of Jesus as the actions and life of God the Son, even if these actions were mediated through the agency of a human subject. Further, he claims, via this interpenetration, God really did share the experiences of Jesus, thus also allowing for us to say in some true sense that God experienced the life of a human being.

A view similar to this is also held by Karl Rahner, widely regarded as the greatest twentieth-century Roman Catholic theologian.[10] According to Rahner, to say that Jesus Christ is God the Son does not mean that there was no human agent or subject of consciousness, since he claims this would subsume Jesus's humanity into his divinity, which is the heresy of the Monosphysites (Dych 2000, 68). Rather, Jesus was a human being who, like us, had a human center of consciousness and action. The way Jesus is different and unique from us in that he definitely, completely, and absolutely accepted God's gift of himself to us, a gift in which we participate "in God's own life through knowledge, freedom and love" (Dych 2000, 71). As Karen Kilby notes, according to Rahner, Jesus is "a human being, one of us, definitively and absolutely accepting God's self-gift" (Kilby 2004, 27). In the language of Rahner, this gift is God's "self-communication" to us. Because of this, it can be truly said that to see Jesus is to see God, since Jesus offers the perfect and unsurpassable human mediation of God. For Rahner, the fact that it is unsurpassable implies that it is not merely a finite mediation of God, since then it could be surpassed (Dych 2000, 74); thus, in Rahner's words, we have to say that God's gift of himself to Jesus "is not only established by God, but it is God himself" (quoted in Dych 2000, 74). According to Rahner, such a complete acceptance of God's gift of himself is God's purpose for all of us, but it is only through Jesus' perfect acceptance of this gift that we are able to come to fully accept it ourselves.

Whatever the merits of this type of view of the Incarnation, it does not pose any obvious problem for multiple incarnations in other races, even an infinite number of other races: taking Ward's rendition of the view, for each race, God could simply unite God's self with an individual in each race from their birth, bringing them into the closest possible union with God's self; or taking Rahner's rendition of the view, a member of each race could perfectly accept God's self-gift, for not only himself or herself, but for all members of the race. This would in no way diminish the fullness of being of God the Son for the same reasons as presented for the two-minds view. Rather, as in the case of the two-minds view, arguably it would enhance God's being.

The Kenotic View

Finally, we come to the so-called kenotic view of the incarnation.[11] This is the only view that poses any obvious philosophical problem for multiple incarnations. This view first arose in the nineteenth century (Swinburne 1994, 230). It says that in taking on a human mind, God the Son emptied himself of the divine attributes, such as omniscience, that conflicted with him being fully human. Thus, it is called the kenotic model, from the Greek word *kenosis,* which means emptying. This word appears in Philippians 2:6–8, known as the *kenosis hymn:* "though he was in the form of God, [he] did not regard equality with God as something to be exploited, but emptied himself, taking the form of a slave, being born in human likeness. And being found in human form, he humbled himself and became obedient to the point of death—even death on a cross" (NRSV).

As Thomas Thompson and Cornelius Plantinga (2006) have argued, the kenotic view of the Incarnation requires the so-called social conception of the Trinity, a model that has recently become popular in Protestant theology and philosophical theology.[12] In this model, God is conceived as three distinct centers of consciousness and agency that are by their essential nature united in the deepest possible way while remaining truly distinct. The kenotic view of the incarnation then makes the additional claim that the center of consciousness and will corresponding to God the Son took on a human body and mind and then subjected himself to the limitations of knowledge and power required by being human.

This view encounters potential problems if one postulates an enormous number of races of VECAs in which God the Son became incarnate.[13] To show why, I begin by arguing that the kenotic view only makes sense if God is in time and there is some sort of 'real becoming' in reality. Views of time committed to real becoming reject the so-called block universe view in which all events past, present and future exist and have equal reality and ontological status. Under the block universe view, past, present, and future only refer to the speaker's temporal relation to an event, and not to any independent ontological status of the event. Consequently, for God, who sees reality just as it truly is, no event is past, present, or future: the election of Ronald Reagan as president of the United States in 1980, for instance, is no more part of the past than what is currently happening today, or what will happen a hundred years from now. Further, nothing ever really changes from the *true* point of view on reality—that is, God's point of view. All events are temporally related to each other as the marks on my desk are spatially related to each other, but without any actual becoming taking place.

Why doesn't the kenotic view make sense under a block universe view? First, the fact that God the Son gives up the divine attributes implies that at one time the Son had these attributes and then at another time he did not. Now, one could try to reconcile this with the block universe view by claiming that temporal indexes apply to the Son in the same way as to human beings. Just as I did not have a beard in 1986 but did in 1998, so one might claim that the Son had all the divine attributes for all times except 0–33 AD. This, however, introduces a temporal indexing and partitioning of properties into the very being of the Son. Consequently the Son's states of consciousness would be partitioned into being states of consciousness at various times; he would not simply possess one total, unified state of consciousness. Thus, for instance, the state of consciousness of the Son at 25 AD would be partitioned off from other states (which are as real) in such a way that the

Son at 25 AD is not aware of the omniscient contents of his consciousness at 36 AD. Furthermore, unless for every incarnation in other worlds, the Son had such partitioned sequences, the Son could not undergo kenosis in that world. Such partitioning, however, threatens to make the kenosis view collapse into the two-minds view, thus losing its distinctiveness.

The kenotic view with multiple incarnations does not run into this partitioning problem under a nonblock universe view with real becoming and God being in time. Since those committed to temporal becoming typically hold that only the present exists, the states of consciousness of other incarnations do not exist at AD 25 and so there is no need for this sort of partitioning off of parts of the Son's consciousness. It does run into other serious problems, however, if there are an enormous number of other incarnations, as I will now argue.

Under a real temporal becoming view, there is an absolute metaphysical distinction between past, present, and future events: present events, for instance, are 'taking place' in an absolute, nonrelational way, whereas past events are already fixed and future events either only exist as possibilities or as having the status of being such that they have 'yet to take place'. Since, on this view, whether an event is taking place is a fact about reality that is not relative to anything else, it would be the same across the races of VECAs: that is, an event is taking place for one race if and only if it is taking place for all races. If there is real becoming, therefore, then at any time t on earth there is some class, C_t, that includes all events that are taking place, and only events that are taking place at t. This class will consist of subclasses according to which race of VECAs they are located at—for example, subclass k will consist of those events taking place in the vicinity of race k. This can then be used to define a universal time across races: an event taking place in the vicinity of race R1 occurs at the same time t as an event in the vicinity of race R2 if they both belong to class C_t.

Now presumably under kenotic Christology, when God the Son was incarnate on earth he had a single unified consciousness that was not simultaneously thinking the thoughts of a member of another alien race. Consequently, the Son could not be simultaneously kenotically incarnated in other races, but could only be sequentially incarnated. Under this view, therefore, if no race has a life span greater than some upper limit, and any effective incarnation takes some minimum time, then if there are too many races of fallen VECAs existing at the same time, there would not be enough time for the Son to become incarnated in most of them. For example, if no race lived longer than a million earth years, and there were a trillion simultaneously existing races, and the minimum time for an incarnation was ten years, then the Son could only be incarnated in less than one in a million races. If there is no overall upper limit to the life span of a race (or minimum time for an incarnation), then it is possible for God the Son to sequentially incarnate: for example, by each time incarnating in a longer-lived race of VECAs, or each time incarnating for a smaller duration of time.

It is difficult to judge the plausibility of this latter scenario, since a universal time does not commit one to the existence of a universal duration. For example, if some other universe had the same atomic structure as ours, the half-life of uranium-235 *as measured by clocks in their world physically identical to ours* would be the same as ours, but it could correspond to one thousand times the half-life of uranium-235 in our universe. Such differences in the rate that identical physical processes occur already are part of our universe: for instance, because of earth's gravitational field, physical processes—such as the decay of uranium-235—occur on earth at a slower rate than in outer space, as predicted by Einstein's general theory of relativity. The most famous example of this is Einstein's twin paradox in which a twin that travels near the speed of light to some planet ten light-years away only experiences a day as having passed upon returning to earth, whereas the twin on earth has experienced twenty years as having elapsed. So, it is at least conceivable that the physical conditions of the races of VECAs could be arranged in such a way that there might not be an upper limit to the life of a race or a lower limit to the length of time an incarnation requires, at least as measured by the clocks of any given race. The event of some race k of VECAs physically like humans beginning to exist, for instance, might take place when the events in our world of October 3, 200 AD take place, and the event of that race going out of existence might take place when the events in our world of October 4, 200 AD take place; yet it is possible that from the perspective of race k, they existed for a million years as measured by clocks in their world physically identical to earth clocks.

In fact, it is conceivable that there could be an infinite number of incarnations that happen sequentially but the sum total of which only take some small finite amount of time. For instance, suppose that the durations of the infinity of different races of VECAs is arranged such that among the incarnations that take place, the longest took 33 earth years, the next longest took 1/2 as long, the next longest 1/4 as long, and so forth *ad infinitum* in a decreasing geometric series. Since the limit of the sum of 1 + 1/2 + 1/4 + 1/8 + . . . *ad infinitum* is 1, the total time the Son would be incarnate for this infinity of races would be 66 earth years, assuming no two incarnations took exactly the same length of time.[14] So, although arguably kenotic Christology does run into significant philosophical problems with there being an enormous number of races of VECAs, these problems might be surmountable with some special assumptions about duration of time across races.

Although I was once attracted to the kenotic view, I now find it less attractive on several grounds. First, it is relatively recent, first arising in the nineteenth century, and so does not have a long tradition in support of it. This should give us some pause in adopting it, though I would not consider this a compelling objection. Second, this view entails that during Jesus's ministry, the Son of God was not upholding the universe, with this function being transferred to the other members of the Trinity. At the very least, this is in tension with passages such as Hebrews 1:3, which says that

God the Son "sustains all things by his powerful word" (NRSV) and Colossians 1:17, which says that "in him all things hold together" (NRSV) Does everything suddenly stop being held together in the Son when he becomes incarnate? Third, as noted earlier, unlike the other views, it commits one to a particular conception of the Trinity—the social conception—a conception that I am unsure is adequate. Finally, arguably it does not really accomplish the kind of solidarity between God and humans that often serves as its major attraction—a God who in some deep way actually experiences the human condition with its vulnerability and pain. In the kenotic view, only God the Son fully shares in human pain and suffering, not God the Father or God the Holy Spirit. The other two members of the Trinity share in our suffering only by proxy—that is, by being united in a deep way with God the Son. Arguably, however, God the Father and the Holy Spirit could have shared in our suffering to the same extent by uniting themselves with the human life-situation, as under the other conceptions of the incarnation. So, when examined carefully, it is unclear whether this major attraction of the view holds up.

NOTES

1. For essays discussing the possibility of the existence of such a multiverse, see Carr (2009).
2. As I define it, the epistemic probability of a proposition A on a proposition B is the degree of support that proposition B gives to proposition A. When B includes all and only the propositions one believes, then the epistemic probability of A on B is the amount by which one should believe A, given one's entire set of beliefs.
3. By being 'fallen', I mean being in a state of needing redemption. In using this term, I am not assuming any original state of righteousness from which we fell.
4. If one responded by claiming that H2 indeed has a prior probability very, very close to one—that is, $P(H2|k^*) \sim 1$—then it follows that it is very unlikely that the Christian faith is true. The argument goes as follows. Let C represent the Christian faith is true. Now, C entails E, and hence $P(C|k^* \ \& \ H2) \leq P(E|k^* \ \& \ H2) = 1/N$. Further, since either H2 or –H2, it follows that if $P(H2|k^*) \sim 1$, then $P(-H2|k^*) \sim 0$. (Here, "–H2" means the negation of H2.) By the probability calculus, $P(C|k^*) = P(C|k^* \ \& \ (H2 \text{ or } -H2)) = P(C|k^* \ \& \ H2) \ P(H2|k^*) + P(C|k^* \ \& \ -H2)P(-H2|k^*)$. Since $P(-H2|k^*) \sim 0$, and for large N, $P(C|k^* \ \& \ H2) \sim 0$, and probabilities cannot be greater than 1, it follows that $P(C|k^*) \sim 0$.
5. Descending Christology and ascending Christology roughly correspond to what has been called 'Christology from above' and 'Christology from below', respectively (see Dych 2000, 67–69).
6. For a more recent discussion of Morris's and Swinburne's views, see Cross (2009).
7. Swinburne (1994, 197) addresses this concern by denying impassibility and claiming that the Fathers of the Church were just wrong in their belief that God could not suffer. Ward's point, however, does not depend on whether or not the doctrine of divine impassibility is correct. His point is that if they were claiming that God the Son was the agent, they certainly would have been

aware of the contradiction. Since presumably they were not making claims that they believed were contradictory, it follows that those who wrote the conciliar statements could not have meant to claim that God the Son was the underlying center of consciousness and agency.

8. According to Swinburne (1994, 197), the claim that Jesus had a human soul should not be taken as meaning that Jesus had a human center of consciousness and action, but rather a human way of acting and thinking: "He [Jesus] can only take on a human soul in the case of a human way of acting and thinking." Since this is not how many theologians following Constantinople interpreted the claim that Jesus had a human soul, Swinburne (1994, 252–53n15) also briefly addresses the history of Medieval interpretations of this claim.

9. John 1:1 states, "In the beginning was the Word, and the Word was with God, and the Word was God" (NRSV); John 1:14 states, "The Word became flesh and lived among us, and we have seen his glory, the glory as of a father's only son, full of grace and truth" (NRSV); and Philippians 2:5–8 says, "Let the same mind be in you that was in Christ Jesus, who, though he was in the form of God, did not regard equality with God as something to be exploited, but emptied himself, taking the form of a slave, being born in human likeness. And being found in human form, he humbled himself and became obedient to the point of death—even death on a cross" (NRSV).

10. According to Swinburne (1994, 228), on the Protestant side Wolfhart Pannenberg was a major near-contemporary theologian who advocated a Christology like this.

11. For an early defense of kenotic view, see Feenstra (1989). For current defenses and critiques of the kenotic view, see the essays in Evans (2006), such as Stephen Davis's (2006) defense of the orthodoxy of the kenotic view. Swinburne (1994, 230–33) also offers a critique of this view. Since, as noted later, the kenotic view is committed to social conception of the Trinity, any critique of the latter will indirectly be a critique of the former.

12. For classic defenses of the social conception of the Trinity, see Plantinga (1989), Brown (1985, 1989), Swinburne (1994, 170–91), and LaCugna (1991). For a critique, see Ward (1996, 321–29) and Tuggy (2003, 168–71).

13. Morris (1986, 183) briefly notes the type of problems presented in the following but does not develop them in any detail.

14. In addition, one might attempt to reconcile the existence of multiple incarnations across races with kenotic Christology by arguing that these other worlds are on different time lines than ours, and hence this problem does not arise. Since real becoming requires a universal time, unless one means by other timelines different durations, other timelines would require adopting a block universe view, thus running into the partitioning problem mentioned earlier.

REFERENCES

Brown, D. 1985. *The Divine Trinity*. London: Duckworth.
Brown, D. 1989. "Trinitarian Personhood and Individuality." In *Trinity, Incarnation, and Atonement: Philosophical and Theological Essays*, edited by R. J. Feenstra and C. Plantinga, 48–78. Notre Dame, IN: University of Notre Dame Press.
Carr, B., ed. 2009. *Universe or Multiverse?* Cambridge: Cambridge University Press.
Collins, R. 2005. "How to Rigorously Define Fine-Tuning." *Philosophia Christi* 7: 382–407.

Cross, R. 2009. "The Incarnation." In *The Oxford Handbook of Philosophical Theology*, edited by T. Flint and M. Rea, 452–75. Oxford: Oxford University Press.

Davis, S. 2006. "Is Kenosis Orthodox?" In *Exploring Kenotic Christology: The Self-Emptying of God*, edited by S. Davis, 112–38. Oxford: Oxford University Press.

Dych, W. 2000. *Karl Rahner*. London: Bloomsbury Academic.

Evans, S., ed. 2006. *Exploring Kenotic Christology: The Self-Emptying of God*. Oxford: Oxford University Press.

Feenstra, R. 1989. "Reconsidering Kenotic Christology." In *Trinity, Incarnation, and Atonement: Philosophical and Theological Essays*, edited by R. J. Feenstra and C. Plantinga, 128–54. Notre Dame, IN: University of Notre Dame Press.

Kilby, K. 2004. *Karl Rahner: Theology and Philosophy*. London: Routledge.

LaCugna, C. M. 1991. *God for Us: The Trinity and Christian Life*. San Francisco: HarperCollins.

Meyer, S. 2010. *Signature in the Cell: DNA and the Evidence for Intelligent Design*. San Francisco: HarperOne.

Morris, T. 1986. *The Logic of God Incarnate*. Ithaca, NY: Cornell University Press.

Morris, T. 1989. "The Metaphysics of God Incarnate." In *Trinity, Incarnation, and Atonement: Philosophical and Theological Essays*, edited by R. J. Feenstra and C. Plantinga, 110–27. Notre Dame, IN: University of Notre Dame Press.

Plantinga, C., Jr. 1989. "Social Trinity and Tritheism." In *Trinity, Incarnation, and Atonement: Philosophical and Theological Essays*, edited by R. J. Feenstra and C. Plantinga, 21–47. Notre Dame, IN: University of Notre Dame Press.

Swinburne, R. 1994. *The Christian God*. Oxford: Clarendon Press.

Tanner, N., ed. and trans. 1990. *Nicaea Ito Lateran V*. Vol. 1 of *Decrees of the Ecumenical Councils*. Washington, DC: Sheed and Ward/Georgetown University Press.

Thompson, T., and C. Plantinga Jr. 2006. "Trinity and Kenosis." In *Exploring Kenotic Christology: The Self-Emptying of God*, edited by S. Davis, 165–69. Oxford: Oxford University Press.

Tuggy, D. 2003. "The Unfinished Business of Trinitarian Theorizing." *Religious Studies* 39: 165–83.

Ward, K. 1994. *Religion and Revelation: A Theology of Revelation in the World's Religions*. Oxford: Clarendon Press.

Ward, K. 1996. *Religion and Creation*. Oxford: Clarendon Press.

12 Incarnation and the Multiverse

Timothy O'Connor and
Philip Woodward

> It has to be said that the Divine Person, over and beyond the human
> nature which He has assumed, can assume another distinct human
> nature.
>
> —Thomas Aquinas, *Summa Theologica*

Traditional Christians affirm the doctrine of the Incarnation—the doctrine
that God the Son, second person of the divine Trinity, became fully human
as the man Jesus of Nazareth while remaining fully divine. The doctrine
developed over the course of the first few centuries of the Christian faith.
During that time, weaker (and less metaphysically puzzling) alternatives
were ruled out by the councils of the Church, alternatives such as that God
only appeared in the form of a human; that Jesus was only an especially
God-conscious human; or that the divinity of God and the humanity of
Jesus were somehow strongly correlated for a time, but not really bound
together in a single individual.

Why have Christians opted for the strongest—and hence least
comprehensible—conception of the Incarnation? The Incarnation is thought
to serve certain divine purposes—purposes that would not have been served
had the divine–human nexus been less intimate than the orthodox position
specifies. At least two such purposes seem relevant.

First, most Christians hold that the Incarnation is *essential to the rescue operation that God brings about in Jesus*. Christians understand the
details of the rescue operation—called 'the Atonement'—in various ways.
On one conception, the death of Jesus expresses a righteous judgment of
and restitution for human sin. In offering that restitution on our behalf,
Jesus makes it possible for us to be forgiven and be restored to fellowship
with God. Jesus could appropriately serve as our representative in this way
only if he became one of us, our elder and blameless 'brother'. On a second
conception, Jesus' life, death, and especially his resurrection (can) liberate
us, individually and communally, from the captivity of sin and its destruc-
tive consequences, including death itself. On still a third conception, Jesus'
living a self-denying, love-filled life and his voluntarily suffering an unjust

death provide the only model for a fully formed human life and unleash a power of love that can transform us. Finally, the Eastern Orthodox teach that the union of the human with the divine in Jesus and his subsequent exaltation pave the way for us to gradually partake of the divine nature. On all these models—and we note, that, importantly, they are not mutually exclusive—we must cooperate with God in some manner for these benefits to flow to us; the chief work of the Atonement is wrought by Jesus, but we must respond to and appropriate it. Further, on all these models, if the union between Jesus' divine nature and human nature is less than complete, a prerequisite for God's purposes for the atonement of humanity would fail to be satisfied. (For some, the necessity here is merely conditional. According to them—Aquinas, for example—while it is 'fitting' that God chose the incarnational path of atonement, he was free to do so in other ways.)[1]

The Incarnation is also thought to serve a second divine purpose: in identifying with human beings in such an intimate way, God thereby *affirms human nature to be of profound intrinsic value.* What is so special about human nature such that God would wish to so identify with it? The creation narrative in the biblical book of Genesis states that humans are divine *ikons,* image-bearers of God. This statement is seen by many theologians as, first, a recognition of certain intrinsic features of human beings, such as our capacity for rationality, for self-awareness, for freedom, and for self-emptying love; and, secondly, as a two-fold gift befitting those same features: the offer of friendship with God and the promise of an eventual, fuller realization of our potential. Indeed, theologians have suggested that the even stronger 'divine-image' language used of Jesus Christ in the New Testament signifies that in the risen Jesus humanity is most fully realized.[2] Our future hope is that we shall be similarly exalted through our identification with him. We shall return to this important theological consideration later.

Now, let us suppose that the basic doctrine of the Incarnation in its ecumenical fundamentals is coherent and let both of the claims just indicated concerning the divine purposes for Incarnation (rescue operation, however understood, and affirmation and future transformation of human nature) be treated as corollaries of it. We suggest that modern scientific understanding of the scope of created reality and plausible theological reflection in a Leibnizian vein both pose a *prima facie* problem for the plausibility of one aspect of the doctrine: its claim of uniqueness.

It turns out that humans inhabit a vanishingly small fraction of known spatiotemporal reality. Might there be creatures elsewhere in our immense cosmos that satisfy the intrinsic conditions for bearing the divine image? This matter is much debated in astrobiology. For all we know, it could be that the probability of the appearance of divine-image bearing (henceforth DIB) creatures is so small that it may take a cosmos 100 billion light years across and 14 billion years old to generate a single DIB species.[3] But, likewise for all we know, the universe might be richly populated with creatures capable of self-awareness, rationality, freedom, love and so forth, to the same or greater degrees than ourselves. And that's just when we contemplate the *confirmed* scope of spatiotemporal reality. Recent, scientifically motivated multiverse hypotheses explode the scale of *contemplated* physical reality to a nigh unimaginable degree. For those who suppose that such cosmological theorizing has a significant measure of empirical support—this, too, is hotly debated—the epistemic likelihood that many other DIB creature kinds exist will be significant, too.

There are also *philosophical-cum-theological* reasons to suppose that reality is a great deal larger than the domain of human observation and influence. Leibniz is surely correct that a being who is necessarily infinitely wise and good will always act for a reason, and indeed (where such is *available*) for the best reason, all things considered. God's actions can bear no trace of value-flouting whimsy or arbitrariness. Among God's actions is the creation of our universe, whose composition is rife with seemingly arbitrary values—the total number of stars, the precise ratio of fundamental particles, the exact speed of light, and so on. It must have been *good* for God to create our universe, else he would not have done so. But it seems that it would have been good also for God to create a universe of a more or less different fundamental character.

There is not space to fully explore this matter, so we will limit ourselves to some brief remarks. Famously, Leibniz held that our *world*—all of actuality: the cosmos, God, and whatever else God might have brought into existence—is the best of all possible worlds. He thought that the infinitely wise God would be able to 'solve for' the optimal balance of good-making features of possible created realities analogously to the way that one may solve for a minimal or maximal value of a curve or size of a region in calculus. In particular, God would solve for the maximin value of a world with endless variety and plenitude that is governed by extremely simple fundamental principles, this global feature being to Leibniz's mind the chief determinant of world perfection. Such variety can be achieved in part by infinite compositional descent with distinct forms at each level (i.e., substances that have ontologically unique parts that have ontologically unique parts that have . . .).

We note that this general approach of seeking optimal balance of goods is consistent with its turning out that a very large (possibly infinite) multiverse figure into the desired solution. And, indeed, there is to our minds a plausible argument from incommensurable goods for such a conclusion (although it is one that Leibniz could not accept). While our universe plausibly is very good in some respects—for example, in orderliness, in beauty, in its capacity to give rise to morally free creatures—at least some of these goods may come at the expense of other possible goods, for example, kinds of structured complexity inconsistent with the kind exhibited in our universe, corresponding kinds of beauty, and creaturely flourishing unsullied by the possibility of moral evil. That is, these other possible goods and some actual goods could not co-exist *within a single universe governed by uniform natural laws;*

consequently, cosmos-building requires trading some such features off against others. If this is so, the question then becomes whether *inconsistent* sets of good-making features are themselves *incommensurable*—whether such sets are incapable of being ranked with respect to overall metaphysical goodness.

Leibniz's negative answer to this question seems to have been heavily determined by his supposing (at least much of the time) that the very abstract and general good of fecundity-from-simplicity is the chief determinant of divine choice.[4] But we doubt that. It seems more likely that more 'local' goods need to be weighed alongside such 'global' goods in determining a universe's value. Such local goods will pertain to some kinds of individuals (including all sentient beings) and their flourishing, the species of which they are instances, and less-than-fully global localities, such as ecosystems. Corresponding to each of these categories, there will be structural goods of various kinds (e.g., involving one or more of the categories of metaphysical, aesthetic, moral, sociopolitical and epistemic). If the possibilities for natural laws and basic kinds of goods vary sufficiently widely, it seems likely that there will be good-making features that cannot sensibly be ranked with respect to each other. And if this is so, it seems further likely that at least some universes of great goodness will be incommensurable, in virtue of exemplifying inconsistent sets of such localized good-making features. The result is that there is a plurality of intrinsically good universe-types; in place of a great chain of (possible) being, there is a great branching tree.[5]

So where does that leave us? We should, we think, go with Leibniz at least to the extent of supposing that God would be *disinclined* to pick one value over another arbitrarily; He would do so only if forced to choose. But, we note, he *needn't* choose between the options, as he might create the best of every class of possible universes whose members are commensurate in value. This collection of top-valued members among value-ordered branches of possible universes would *collectively* constitute the best possible world. Quite possibly, many of these universes possess the value of containing DIB creatures. (Leibniz would here object that such a multiverse would ruin the organic value of reality as a whole. We doubt that it is sensible to apply the notion of organic value to collections of almost completely disconnected totalities, but even if it is, it seems plausible in the envisioned scenario to suppose that this drawback is amply outweighed by the goods secured in realizing all of the best universes of their value-kind. And the selection of this particular array would hardly be purely arbitrary, and so might admit of a sort of organic unity applicable to collections of universes, if such there be.)

But many depart more radically from Leibniz by rejecting his assumption that there *is* a best possible world. Suppose that he was so mistaken. Perhaps, for each (or some) of the valued-ordered branches of commensurable universe kinds, there is no *top* value (corresponding to one or more of the kinds). We think that this scenario, too, points in the direction of a multiverse, indeed of an infinitely membered one. For it is hard to credit the thought that a perfectly wise being of limitless power, contemplating each of the infinitely ascending branches of the creative possibilities, should just arbitrarily pick one from each, fully aware that whichever one he picks, no matter how far up the scale it resides, there are others of enormously greater value than it. Again, if He had no choice but to make such an arbitrary selection, we would suppose—unlike Leibniz, who was irrevocably committed to the Principle of Sufficient Reason—that he might well do so.[6] But He did have another, less arbitrary choice. For he could choose an appropriate threshold of goodness and create every one of the infinitely many universes above it (or every other one, or every millionth one, or . . .) In this case, almost certainly, infinitely many of these universes possess the value of containing DIB creatures.[7] It is hard to say what would be an appropriate, nonarbitrary threshold of goodness. The most obvious candidate is any universe with on-balance positive value. But this fails to take account of more 'local' considerations, such as passing over universes that involve intense suffering of persons without even the prospect of their acquiring redemptive significance. Once one begins to consider plausible such constraints, epistemic modesty seems the order of the day: we, severely cognitively limited creatures that we are, just cannot say where the line would be drawn by a morally and cognitively perfect being.[8]

For these reasons, Christians have significant (though by no means definitive) scientific and theological reasons to leave open the possibility that there are other DIB creatures in existence. But if so, it would seem that the divine purposes behind the human Incarnation would also apply to these other beings: supposing any of them were in need of rescuing of the sort that Christians believe we are in need of, taking on their natures would presumably be a prerequisite for such saving work among them. And even if no rescuing were needed, the second divine purpose—identification with the lives and experiences of DIB creatures—would apply anyway.

In response to this suggestion, a Christian might say that God's human Incarnation in Jesus of Nazareth serves both these purposes for *all* DIB creatures. After all, human persons vary considerably, yet God's Incarnation as the particular first-century Palestinian man Jesus of Nazareth is thought to serve God's restorative and identifying purposes for all of us. Why not for all DIB creatures, human and nonhuman alike?

There are a couple of reasons to find this response unsatisfying. First of all, it suggests that we humans won an Incarnational lottery—that we alone, for no apparent reason—were chosen as the recipients of God's incarnational act. Here again, Leibnizian worries about arbitrariness loom. Why would God choose us rather some other DIB species among which to be incarnated?[9]

A second problem for the suggestion that God's Incarnation as the human Jesus serves God's purposes for all DIB creatures has to do with an implied epistemic ignorance, of Jesus' life and work, by these creatures. While it may be that God's purposes for other DIB creatures can be served without their knowing about it, Christian devotional practice reflects the view that an eventual *awareness* of God's redemptive work is a great good for us, a source of comfort[10] and joy.[11] Further, inasmuch as God's redemptive work includes the formation of a community of creatures in covenant relationship with God, and inasmuch as that community was inaugurated by and remains formed around the Incarnate Son (as is implied by the biblical language of the Church as a 'body' whose 'head' is Christ), it would appear that redemption cannot be complete for a DIB creature who lacked awareness of the Incarnation and connectedness to the community it inaugurated. For it is plausible that one has not been fully folded into a community unless she is aware of that community's existence and *raison d'etre.*

Well, maybe you buy the foregoing reasoning, and maybe you don't. But even if there happen not to be any DIB creatures save human beings, or there are, but the life, death, and resurrection of Jesus serves God's purposes for all of them, there is an underlying metaphysical issue worth exploring: whether or not it is within the scope of an omnipotent being's power to take on more than one DIB nature. Might God have been multiply incarnated? If so, how might this work? With Aquinas (in this essay's epigraph), we hold that a viable metaphysics of the Incarnation has the consequence that multiple Incarnations are indeed *possible.* Here we can but sketch a way of modeling this possibility and consider a few objections.

HOW TO BE AN INCARNATE DEITY

As with the doctrine of Atonement, so with the doctrine of Incarnation: all traditional Christians affirm it, but there is much disagreement concerning how to lay a conceptual foundation for beginning to understand it. The core thesis is that the second person of the Trinity, God the Son, took on a full human 'nature', so that He became a single person having two natures, human and divine. *Prima facie,* this is incoherent, as at least certain of the essential properties of divinity and humanity seem incompatible. A number of theories have been offered to show that first appearances are deceiving in this case. The theological and philosophical issues they raise and difficulties they face are complex, and we shall not try to survey them here.[12] Instead, we will indicate our preferred theory, and that only briefly. While this theory certainly does not dispel all mystery surrounding the Incarnational doctrine, it does provide a model that doesn't have incoherence on its face. We will then deploy it to consider the possibility of many-natured incarnation.

The view that we propose is the *compositional theory*—or rather, a particular version within the family of compositional theories. The core idea here is that, in taking on a full human nature (mind and body), the single divine person becomes a *composite* thing or substance. He is the self-same person, retaining all the omni-attributes of divinity, but now has and expresses two natures, each of which are distinct components of his being.[13]

Where we go from the core idea in developing the compositional theory depends in part on how we think of human persons. We shall follow neither the many Christians who thought or think of humans as immaterial substances, nor the medieval Aristotelians such as Aquinas who thought of them as matter-form compounds. Instead, we hold that we (merely) human persons are wholly materially composed individuals who have a kind of unity not had by garden-variety material composites. This unity is conferred by our having strongly emergent mental capacities and properties, which are metaphysically basic—*not* physically realized—features that make a (non-redundant) causal difference to the way the world unfolds. (In the familiar older lingo, our view is a substance monism about the human person conjoined with a strong form of nonepiphenomenal property dualism.)[14]

On our preferred version of compositionalism, when God the Son became incarnate, he simultaneously created and absorbed *into himself* a human embryo which, as it matured, manifested increasingly rich mental capacities and properties.[15] That developing embryo-fetus-newborn-youth-adult was (and eternally is) not a distinct person from God the Son, co-Trinitarian-participant in the creation of the world. It was (and is) an instance of human nature, a living, fully intact human body, but one that is not, *in itself*, a person at all; it is a part, the human part, of the one person, God the Son, latterly known as Jesus Christ in virtue of the incarnational event.

Now, an important task for any would-be compositional account of the Incarnation is to specify the relation that holds between the components of this divine–human being, such that they are substantially unified and together constitute a single person. In agreement with most orthodox theologians, we doubt that this task can be fully accomplished: we human beings lack the conceptual resources to fully penetrate the mystery of the Incarnation. But some things can be said that go a certain distance.

One adequacy constraint on such an account is that it makes clear why the human component of the divine–human individual does not constitute a purely human person in its own right. To that end, we suggest that persons are individuated by their being both *a center of subjectivity* and *the wellspring of the acts* they perform. In other words, sameness of person entails sameness of subject and sameness of agent. Typically, an instance of human nature will include, in itself, a proprietary center of subjectivity and agency; that is, a properly formed and functioning human body is sufficient for the emergence of an autonomous, experiencing *subject* and *agent* at the center of a dynamic phenomenal/intentional manifold. But were the human nature of Jesus to include a proprietary human center of subjectivity and/or agency, we would have on our hands a *complete (solely human) person,* or so it seems to us.

Here's what we propose. When God the Son took on a human body as a part, the emergence base for that human body's mental states was expanded. The base then included not only the types of causal powers that would ordinarily be sufficient to generate an experiencing subject and agent at the center of a dynamic phenomenal/intentional manifold. It also contained divine causal powers that *masked* the causal powers responsible for the emergence of a proprietary human subject. Conscious mental states nevertheless emerged, but absent a proprietary *human* subject, they emerged as mental states *of the larger individual,* the divine–human composite.[16]

Yet the Christological creed of Chalcedon also teaches that Jesus had distinct and 'unmingled' human and divine 'intellects' and 'wills'. To square our proposal with this creedal declaration, we suggest that the one person, the Son, somehow operates *through* his human intellect—experiencing as subject the purely human phenomenal/intentional manifold—and through his human will—*initiating* in some distinctive way the human acts of will that operate in the characteristic manner of human action.[17]

Pulling the threads together, the eternal Son of God is a divine person having essentially the divine omni-attributes. At a point in time, he co-created and in a mysterious manner grafted into himself a living human body, such that it was from its inception *his* body. Like other properly formed living human bodies, this body also exhibited the attributes of human, finite *person*hood; it was an unfolding sphere of changing, finite, perspectival phenomenal and intentional states (intellect) and of limited agency (will). But while there are two sets of distinctively personal capacities of intellect and will, human and divine, there is but one person. There is a single locus of subjectivity and agency, anchored in the divine mind, which is manifested in part through the embodied human mental capacities of Jesus of Nazareth. In this way, Jesus is fully human while being (metaphysically) unique among humans.[18]

HOW TO BE A MULTIPLY INCARNATE DEITY

Consider the person known as Jesus of Nazareth on earth and as Joshua of Namoth on Gliese 581g (thought to be the nearest planet outside our solar system that falls within the 'habitable zone' of its solar system). And consider the suggestion that these apparently distinct persons are in fact the very same divine (multiply creaturely incarnated) person. This requires the possibility that one person can occupy two widely separated spatial regions. But note that it is not a case of multilocation of *bodies,* whereby one wholly material object wholly occupies more than one spatial region. (And that's a good thing, for this kind of multilocation, though toyed with by some recent metaphysicians, is a highly problematic notion.)[19] For the two *bodies* of Jesus/Joshua are distinct objects, parts of the one person who lives through them. As we wrote, on the Incarnational picture we propose, these bodies, instances of human and Gliesian nature having mental as well as physical attributes, are inherently dependent entities, not proper substances in their own right. But insofar as we think of them in isolation from the one individual they partly compose, they are wholly distinct. To whom/what, then, does Peter refer when he points and says to John, "There is Jesus"?

We take it that he refers to the *person* Jesus. So, if multiple incarnations are actual, he (unknowingly) refers to a person who also, perhaps simultaneously, occupies a planet far, far away. And if he says, "There is the body of Jesus", he makes (on a natural disambiguation of what he says) a mistaken assumption of uniqueness. For the person Jesus has more than one body. Now, if he were philosophically savvier than we have reason to suspect the uneducated fisherman from Galilee really was, he could say truly, "There, and only there, is the *human* body of Jesus".

Now, you might sense a more troublesome oddity when we turn from the body to the mind of Jesus. If Jesus of Nazareth is the very same person as Joshua of Namoth, the thought goes, then Jesus' mental states would seem to be very confused! He would be thinking, for example, "John is my beloved disciple" and "Giles is my beloved disciple". But this thought itself rests on a confusion concerning the doctrine of Incarnation. Jesus is the Incarnate Son of God. He has a fully divine and fully human mind, and these are distinct (albeit overlapping) ranges of thought of one person.[20] The restricted human mental life of Jesus will have no access to thoughts in the Gliesian mind, and vice versa. The human 'mind' of Jesus will presumably not even include awareness that he is incarnated on Gliese 581g. But the eternal Son of God is, in his divine mind, fully and simultaneously aware of all the thoughts flitting through both of the creaturely minds associated with the names 'Jesus' and 'Joshua'—that is, *his* creaturely minds. And this is just a special case—a special kind of 'inside' knowledge, owing to his being incarnated as Jesus and Joshua—of his knowledge of *all* creaturely thoughts in Creation.[21] There is, we suggested, but one center of subjectivity in this multiply incarnated divine person. (We might think of the divine mind's awareness of the limited creaturely minds of his incarnations by a very loose analogy to our own awareness of the distinct deliverances of multiple sense modalities, centered in a single subjectivity.)

A THEOLOGICAL WORRY

According to the current proposal, if the Son of God can take on a human body/mind as a part of himself, he can take on (and perhaps has taken on) the natures of many, and potentially infinitely many, other DIB species, without its being the case that any physical thing is wholly multilocated (throughout a single universe or among many) and without fragmentation of the divine mind, which serves as the center of subjectivity and agential control of the creaturely minds resulting from the many incarnations. Even if all this is granted, one might object that our proposal generates *theological* problems. We will confine our attention to a position parallel to Aquinas's on the eternity of the world: while it is metaphysically possible, the *actuality* of multiple incarnations is incompatible with what Christian Revelation teaches.

A number of New Testament passages seem to imply that God's redemptive purposes *for all of creation* are served by the life, death and resurrection of Jesus—that is, by the actions performed by the Son of God through his human nature. Says the author of Colossians: "For God was pleased to have all his fullness dwell in [Christ], and through him to reconcile to himself all things, whether things on earth or things in heaven, by making peace through his blood, shed on the cross" (Colossians 1:19–20). And the writer of Ephesians adds that God's will is "to bring unity to all things in heaven and on earth under Christ" (Ephesians 1:10). But the cohesiveness and comprehensiveness of Christ's redemptive work that these passages assert is undermined by our proposal (or so goes the objection). If there are many DIB species, and the Son of God redeems them and identifies with them by taking up each of their natures individually, then the acts of Jesus recorded in the New Testament do *not* serve to reconcile to God all things.

At one level, this worry can be dealt with pretty straightforwardly. Remember that the acts of Jesus recorded in the New Testament, and the acts of Joshua recorded in a Gliesian text (that is, alas, unavailable to us) are not acts of distinct persons but acts of one and the same person, the Son of God. So on the contemplated multi-incarnational picture, it is indeed through Christ that God reconciles all things to himself, just not exclusively by those of his creaturely actions that are recorded in the New Testament. But it is entirely fitting that the human authors of these scriptural texts would know nothing of Christ's actions in distant galaxies or causally isolated universes. So our proposal does nothing to contradict what we take these passages to be saying, viz. that Christ's incarnate acts *in toto* make redemption possible for all reality, not just for human creatures.

However, there is a deeper and more interesting problem that these New Testament passages raise for our proposal, a problem regarding the eschatological picture that they suggest. The passages are commonly read as suggesting that Christ's work is necessary not only to *redeem* all things but also in an important way to *unify* all things, where this is taken to mean that it will ultimately usher in a harmonious and profoundly united *community* of all DIB creatures, under one authority, the Son of God.

As we see it, however, deep community seems possible only among creatures of broadly similar natures—who have broadly similar needs, who flourish in broadly similar environments, who can form relationships with one another, who can successfully communicate with each other, and so on. So given that DIB species would presumably *not* all share sufficiently similar natures, it is hard to see how deep community among all DIB creatures could be possible. And supposing that God the Son is incarnate in a multiplicity of DIB species, prospects for the envisioned *form* of unification look even stranger. Each species will have known God the Son in the 'dress' of its own nature. In *which* of Christ's creaturely natures would he present himself to a unified community of radically diverse creatures? Any choice would be arbitrary. One might urge that the Son of God would not need to choose, because, in the eschaton, he will be known by creatures solely and directly through his Divine Nature, without the mediation of any creaturely nature. On this scenario, creatures will be transfigured in such a way that direct, spiritual encounter with God is possible. If so, perhaps this transfiguration will also serve to overcome differences between DIB natures such that a unified community is in fact possible. The trouble with this option is that it seems to violate one of the purposes that motivated God's becoming incarnate in the first place, namely, the permanent identification with and eschatological perfection of creaturely natures. If Barth and other theologians are correct that the incarnate Jesus Christ is 'the real man', the fullest realization *of humanity*, then by parity we would expect a Gliesian incarnation, a fullest realization of *that* nature. The only apparent way that this consequence might be avoided is to assume that humans and Gliesians alike are to be transformed into something unrecognizable as distinctively human or Gliesian—a generically DIB nature. However, it seems more in keeping with the implicit theology of the New Testament that redeemed creation maintains its diversity.[22] So we propose instead that distinct DIB species, if such there be, retain their distinctiveness. And we contend that it is consistent with the New Testament passages just cited to anticipate deep unity *within each community of DIB creatures*, with Christ as Lord of all such communities, and with each community retaining its God-affirmed peculiarity even as they participate in a common goal of union with God.

CONCLUSION

To recap: on our model, the divine Son's becoming incarnate is a matter of his becoming composite by taking on as a part an instance of a creaturely nature, complete with its ontologically emergent mental states but without a proprietary center of subjectivity and agency. He can then act through this creaturely nature in the manner and to the extent that the work of the Atonement requires. And he can repeat this process as many times as there are populations of creatures that bear the divine image—even if (as seems reasonably likely to us) there are infinitely many such populations. One divine Son acts among and on behalf of DIB creatures in the many DIB populations by being incarnated in and acting through an instance of each nature.

Given the number of contentious speculative matters on which this conclusion rests, it is fitting to end on a note of epistemic modesty. Throughout our discussion, we have touched on a number of topics central to Christian faith: God's rationality and will, his reasons for creating, his relationship to those of his creatures that bear his image, his redemptive purposes and actions, and so forth. Not only do we presently lack the capacity to penetrate these mysteries, but we have good reason to disclaim our ever acquiring such a capacity. Nevertheless, we take it that our discussion demonstrates that the doctrine of the Incarnation is consistent both with our best philosophical theory of human nature and with scientifically and theologically motivated multiverse hypotheses—and this much more general thesis is itself a substantive philosophical conclusion.[23]

NOTES

1. Of course, it may be that certain theories of the Atonement do not adequately motivate the Incarnation, despite what their proponents contend. If so, that would be a reason to reject the sufficiency of such theories, since an adequacy constraint on them is that they explain why God became incarnate.
2. For an influential development of this suggestion, see Karl Barth (2010, vol. III, part 2).
3. The question hangs in part on the necessary conditions for life, something that is still not fully understood. Even on our own planet, we are finding life-forms ('extremophiles') in conditions that had been thought to preclude life. On the other hand, the seemingly unrelated matter of an active plate tectonic system—which is also responsible for earthquakes and volcanoes—appears to be essential for recycling elements of the atmosphere and regulating the temperature of any life-sustaining planet. It is for want of such an ocean-based system that the planet Venus cannot sustain life, even though it lies within the "habitable zone" around our sun (see Kastings 1996).
4. For discussion of Leibniz on this point, see Wilson (1983).
5. This position is developed in O'Connor (2008, Chapter 5).
6. And also unlike William Rowe (2004), who thinks this scenario points to an a priori argument for atheism. For a reply, see O'Connor (2005).
7. This line of argument is developed in O'Connor (2008, Chapter 5).
8. Which is not to say that we cannot say anything at all. We are inclined to assume, for example, that a perfect Creator would not be motivated to create qualitatively duplicate worlds or worlds which are only trivial variants on other worlds, with no significant difference of *type*. Here, the artisanal image of the Creator looms large in our thinking. Is it only for want of time and other resources that a human artisan, having created an exquisitely beautiful statue, is not strongly motivated to reproduce it? We judge not. Creative fecundity is best measured in types, not tokens. (And note that once duplication is on the table, there is no satisfiable limit, since there is no highest transfinite cardinal.)
9. Leibniz (1952) tacitly acknowledges something like this worry in Part I, Section 18 of the *Theodicy*, when he criticizes an unnamed proponent of a rationalist, 'astronomical theology': "It does not appear that there is *one* principal place in the known universe deserving preference to the rest to be the seat of the eldest of created beings; and the sun of our system is certainly not it". We note, however, that on this particular point, Christian theology teaches that the particulars of Christ's Incarnation—a lowly birth in a cultural backwater—were, despite natural expectations, particularly *fitting* circumstances for the one "who came, not to be served, but to serve" (see Matthew 20:28 and Mark 10:45), and to be an example thereby to all of us.

   Not all philosophers of religion are terribly troubled by this sort of divine arbitrariness. Robert Adams (1972) contends that there is no moral *obligation* to create the best *and* that a choice by God of less than the best can be adequately accounted for in terms of divine grace, a disposition to love independent of the value or merit of that which is loved. And Michael Rea (2011) has recently suggested that the oft-troubling fact of divine silence vis-à-vis his human creatures may reflect in part God's personality, His preferred mode of interaction, rather than anything about the extent of His concern or love for us. We may have these thinkers wrong, but they seem to be suggesting that there may be idiosyncracies to God's personality, characteristics that have no integral connection to God's other omni-attributes. For our part, we can't attribute idiosyncracy and the contingency that seems to flow from it to God's character, given the necessity of his existence and his bearing the traditional omni-attributes.

10. Hebrews 4:15–16 says that "we do not have a high priest who is unable to sympathize with our weaknesses . . . let us then approach the throne of grace with confidence".

11. 1 John 3:1–2 says: "How great is the love the Father has lavished on us, that we should be called children of God! And that is what we are! . . . what we will be has not yet been made known. But we know that when he appears, we shall be like him".

12. The reader seeking a philosophically-sensitive discussion of these theories may consult Cross (2009). For a fuller look at specific options, see the nice collection of essays in Marmodoro and Hill (2011).

13. As you'd expect, there is a thorny issue here connecting this picture of the Incarnation to the doctrine of the Trinity, according to which God the Son is a person who is *of one substance with* the Father and the Holy Spirit. It might seem that the divine-human *composite substance* that, on our account, is the incarnate Jesus Christ could not be of one substance with the other, wholly immaterial persons of the Trinity. While we will not try to finesse that puzzle here, we note that we are inclined to agree with Leibniz that avoiding contradiction requires us to reject the common scholastic view of the communicability of properties between the two natures. But see Stump (2002, 206–7).

14. It's interesting to observe that, in his late correspondence with Des Bosses, Leibniz saw a problem for his mere aggregation view of the human body in application to the Incarnation. If Christ is himself to be a true unity and not a mere aggregation, there needs to be a 'substantial bond' (*vinculum substantiale*) *within* his human nature, something more than his official picture of a colony of monads and their modifications allows. We thank Maria Rosa Antognazza for calling our attention to this discussion, which is available in Look and Rutherford (2007).

15. In Thomas Flint's terminology, our is a 'model T' rather than 'model A' version of compositionalism (Flint 2011).

16. We thank Dean Zimmerman for helpful discussion on this point. Also, it is hard to see why the purely human nature of the Incarnate Son should not be separable from the composite individual—Incarnation does not seem to entail the temporal eternality of Incarnation, even if it is so in fact. What then would be the ontological status of Jesus' particular human nature, were the Son of God to sever ties with it while continuing to sustain it in existence? It seems that this would entail the appearance of a new, purely human individual with merely quasi-memories.

17. We owe this suggestion to Brian Leftow (personal correspondence).

18. As an aside, we suggest that it is worthwhile to think through this proposal by considering whether God could create a *purely creaturely* dual-nature person. (Scenario 1: equal natures, such as human-human. Scenario 2: unequal natures, such as human-'hobbit'.) We are not aware of anyone discussing such non-divine dual-nature scenarios. Our inclination is to suppose that only an omniscient mind could subsume a second nature without massive psychological fragmentation.

19. See Kleinschmidt (2011) for discussion of ways that multilocation would violate compelling axioms of mereology.
20. Compare Thomas Morris's notion of an "asymmetric access relation" between the divine and human minds of the Son (Morris 1986, 103ff.).
21. This matter of perspectival knowledge of course raises questions concerning the nature of omniscience, but we cannot address them here.
22. In Catholic and Eastern Orthodox theology, the term 'communion of saints' is taken to encompass not only redeemed human beings but also the angels with whom they will join in the eternal worship of God. We take this teaching to be consistent with the points we are advancing here, namely, that (1) there is a profound, valuable form of experienced community had by and necessarily restricted to conspecifics, whether they be human, Gliesian, or whatever, and (2) that there is a value to human beings in knowing God via his becoming one of us that is not realized by angelic beings and would not be realized by other DIB creatures apart from incarnation in their natures, and, finally and similarly, (3) that the perfecting of human nature in Jesus's human incarnation would not carry over to other DIB natures apart from parallel incarnations.
23. Versions of this essay were delivered at the following conferences: First Midwest Annual Workshop in Metaphysics, St. Louis University, October 19–20, 2012; a conference on Leibniz's *Theodicy* in Lisbon, October 25–27, 2012; a workshop on God, Time, and Eternity at Queen's University, Belfast, December 11–12, 2012; and at the God and the Multiverse Workshop at Ryerson University, February 15–16, 2013. The penultimate draft was also discussed at a philosophy of religion reading group at Oriel College, Oxford. We thank the audiences at these events for a good deal of constructive feedback. We would like especially to thank Charity Anderson, Maria Rosa Antognazza, Matthew Benton, Jeff Brower, Robin Collins, Alicia Finch, Hud Hudson, Klaas Kraay, Brian Leftow, Tim Mawson, Jeffrey McDonough, Timothy Pawl, Eleonore Stump, Richard Swinburne, Peter van Inwagen, Catherine Wilson, and Dean Zimmerman.

REFERENCES

Adams, R. 1972. "Must God Create the Best?" *Philosophical Review* 81: 317–32.
Barth, K. 2010 (originally published 1957). *Church Dogmatics*, edited by G. W. Bromily and T. F. Torrance. Translated by A. T. Mackay and T. H. L. Parker. Louisville, KY: Westminster John Knox Press.
Cross, R. 2009. "Incarnation." In *The Oxford Handbook of Philosophical Theology*, edited by T. Flint and M. Rea, 452–74. Oxford: Oxford University Press.
Flint, T. 2011. "Should Concretists Part with Mereological Models of the Incarnation?" In *The Metaphysics of the Incarnation*, edited by A. Marmodoro and J. Hill, 67–87. Oxford: Oxford University Press.
Kastings, J. F. 1996. "Planetary Atmosphere Evolution: Do Other Habitable Planets Exist and Can We Detect Them?" *Astrophysics and Space Science* 241: 3–24.
Kleinschmidt, S. 2011. "Multilocation and Mereology." *Philosophical Perspectives* 25: 253–76.
Leibniz, G. W. 1952. *Theodicy*, edited by A. Farrer and translated by E. M. Huggard. New Haven, CT: Yale University Press.
Look, B., and D. Rutherford, eds. 2007. *The Leibniz–Des Bosses Correspondence.* New Haven, CT: Yale University Press.

Marmodoro, A., and J. Hill, eds. 2011. *The Metaphysics of the Incarnation*. Oxford: Oxford University Press.

Morris, T. 1986. *The Logic of God Incarnate*. Ithaca, NY: Cornell University Press.

O'Connor, T. 2005. "Review of William Rowe's *Can God Be Free?*" *Notre Dame Philosophical Reviews*, April 8.

O'Connor, T. 2008. *Theism and Ultimate Explanation*. Oxford: Blackwell.

Rea, M. 2011. "Divine Hiddenness, Divine Silence." In *Philosophy of Religion: An Anthology*, 6th ed., edited by L. Pojman and M. Rea, 266–75. Boston: Wadsworth/Cengage.

Rowe, W. 2004. *Can God Be Free?* New York: Oxford University Press.

Stump, E. 2002. "Aquinas' Metaphysics of the Incarnation." In *The Incarnation*, edited by S. T. Davis, D. Kendall, and G. O'Collins, 197–218. Oxford: Oxford University Press.

Wilson, C. 1983. "Leibnizian Optimism." *Journal of Philosophy* 80: 765–83.

Contributors

Michael Almeida is professor of philosophy and chair of the Department of Philosophy and Classics at the University of Texas at San Antonio.

Robin Collins is distinguished professor of philosophy and chair of the Department of Philosophy at Messiah College, Grantham, Pennsylvania.

Peter Forrest is Professor Emeritus of Philosophy, University of New England, New South Wales, Australia.

Jeremy Gwiazda is a financial operations instructor at a nonprofit organization (Year Up) in Boston, Massachusetts. He completed his PhD in philosophy at the Graduate Center of the City University of New York in 2009.

Klaas J. Kraay is associate professor of philosophy at Ryerson University, Toronto, Ontario, Canada.

John Leslie is University Professor Emeritus, University of Guelph, Ontario, and a Fellow of the Royal Society of Canada.

Robert B. Mann is professor of physics and applied mathematics in the Department of Physics and Astronomy at the University of Waterloo, Ontario, Canada.

Jason L. Megill has been an instructor of philosophy at the University of Colorado, Boulder, and at Old Dominion University, Norfolk, Virginia, and an assistant professor at Carroll College, Helena, Montana. He completed his PhD in philosophy at the University of Virginia in 2008.

Yujin Nagasawa is professor of philosophy in the School of Philosophy, Theology, and Religion at the University of Birmingham, United Kingdom.

Timothy O'Connor is professor of philosophy at Indiana University, Bloomington.

Don N. Page is professor in the Department of Physics at the University of Alberta and a Fellow of the Royal Society of Canada.

Michael Schrynemakers is an instructor of philosophy at St. John's University, Queens, New York, and at Sacred Heart University, Fairfield, Connecticut. He completed his PhD in philosophy at the Graduate Center of the City University of New York in 2013.

Donald A. Turner is professor of philosophy at Hillsdale College, Hillsdale, Michigan.

Philip Woodward is a doctoral candidate in philosophy at Indiana University, Bloomington.

Index

abstract objects 4, 12, 71, 92–5, 104–6, 112, 130; *see also* concrete objects
afterlife 81–2, 89, 123–4, 213
Aristotle/Aristotelian 2, 25, 26, 71
atheism, arguments for 7–9, 12–13, 131–4, 136–45; *see also* evil; problem of evil; suffering
atonement 227–8, 232, 237–8
axiology 6–7, 9, 14, 17, 130–47, 188, 199–200, 203–6

beneficence 13, 153–4, 157
Big Bang 2, 3, 28–9, 34, 36, 40, 66, 84, 86, 108, 184, 192, 205
biophilic/life-permitting characteristics 3, 11, 29–30, 33–41, 195–7; *see also* fine-tuning argument
block universe 62, 87, 198, 221, 225
Boltzmann brains 11, 36–41, 50, 58

Cantor/Cantorian infinity 13, 162–73
cardinality 74, 88, 115, 137, 164, 238
concrete objects 4, 12, 14, 46, 58, 92–112, 116, 137, 162–3, 202; *see also* abstract objects
Copernicanism 31, 41
Copernican Principle 26–8, 96
Copernican Revolution 2, 25, 26
Copernicus 18, 26
counterfactuals of freedom 16, 121–122
counterpart 76–9, 83, 94, 115–16, 118, 143, 185
creation *ex nihilo* 38, 81–5

determinism 67–70, 196, 205
divine attributes: benevolence 11, 38, 45, 52, 55, 138, 180–1, 189–90, 197, 205; *see also* divine attributes, moral perfection/perfect goodness; eternal 119, 181, 203, 217, 233–5, 239; impassible 219, 224; independence 181, 190; moral perfection/perfect goodness 4–5, 7–9, 12, 16, 81, 89, 114, 121–4, 132–3, 150–2, 179–80, 203, 212–14; *see also* divine attributes, benevolence; necessary existence 4, 16, 129–30, 150, 181–5, 203, 239; omnipotence/perfect power 4–7, 9, 11, 12, 16, 38, 45, 52, 80, 82–3, 89, 114–15, 121, 131–4, 136–7, 139, 141–3, 145, 150, 152, 179–81, 189–90, 197, 203, 217, 221, 224, 231–2; omni-presence 181, 217; omniscience/perfect knowledge 4, 7, 9–12, 16, 38, 45, 52–3, 55, 87, 89, 132, 150, 152, 180–1, 189–90, 217, 220, 222, 239–40; self-existence 183, 185, 189; simplicity 81, 89; unsurpassability 4, 8, 10, 130–47, 181, 220; worshipworthy 13, 123, 179–80, 185–6, 189, 190, 240
duplicate/duplication 10–12, 36–7, 98–100, 111, 114–15, 142–59, 238

Einstein 45, 52, 58, 194, 198, 205, 223
elegance/elegant 11, 38–40, 45, 50–8, 79, 129, 195
ersatzism 71, 93–6, 102–12
Everett interpretation of quantum mechanics 3, 11, 36, 47–50, 54–8, 66, 69, 107, 109, 192–3

evil. gratuitous 12, 108, 120, 130, 141–7 158–9; horrendous 154, 156, 188–9; moral 144–5, 153, 213, 229; natural 55, 144–6, 158–9; *see also* atheism, arguments for; problem of evil; suffering
evolution 27–30, 37, 45, 109, 194, 212, 214

fine-tuning argument 3, 33, 65–6, 194–6; *see also* biophilic/life-permitting characteristics
freedom: divine 7, 11, 17, 122–3; compatibilist 66–7, 150, 152, 155, 160, 205; incompatibilist 67; libertarian 5, 122, 150, 152, 156; moral 13, 141, 144–5

Galileo 2, 26
geocentrism 2, 25–7, 30–1, 40
gravity 26, 29–30, 32, 34, 45, 52, 58, 85, 108, 192, 195, 223

heliocentrism 2, 18, 26
Hubble, Edwin 3
Hubble volume 36
hyperspace 11, 18, 61–5, 70–89, 122

identity 10, 64, 76–7, 89, 98, 110–11, 114–19, 135, 137–47, 177–85, 190, 202, 223
Identity of Indiscernibles 10, 84, 111, 114–15, 198, 202
incarnation, models of: compositional 232–8; enhypostatic 217, 219–20; kenotic 211, 217, 220–5; two-minds 217–19
incommensurability 229–30
incomparability 7–8
indeterminacy 66–8, 71–6, 84–7, 129, 131, 147, 196–7
inflation 1–3, 26, 34, 36, 38–9, 192–3, 195–6, 211
injustice 39, 123, 154, 156
island universes 118–19, 151, 182–3

Jesus Christ 14, 18, 33, 54, 57, 215–20, 223, 225, 227–8, 231–40
justice/injustice 13, 39, 123, 153–7, 160

Kant, Immanuel 2, 83, 153

Leibniz, Gottfreid 7, 18, 84, 107, 111, 180, 228–31, 238–40
Lewis, David 4, 12, 13, 16, 62–5, 69–71, 78, 86–7, 92–4, 107–9, 158–60, 182–8, 201; *see also* modal realism
Linde, Andrei 1–3, 34, 39

many-worlds interpretation of quantum mechanics. *See* Everett interpretation
mediocrity, principle of 27–8, 34–41
modal collapse 7, 11, 147
modal realism 12, 13, 16, 65, 78, 86, 92–112, 116, 118, 121, 178, 181–83, 185, 187, 190, 201–2; *see also* Lewis, David
Molinism 84, 121

naturalism 31–2, 35, 62
nominalism 94

objects. *See* abstract objects; concrete objects
Occam's razor 45, 46, 50, 81
ontological argument 183, 189

panentheism 178, 190
pantheism 13–14, 177–91, 197–205
physicalism 62
Plantinga, Alvin 4, 5, 16, 93, 104, 158–160
Plato/Platonic 2, 25, 87, 199–206
polytheism 83
predestination 84
Principle of Alternate Possibilities 160
Principle of Plenitude 94, 130, 229
Principle of Recombination 94, 150
probability: epistemic 213, 214, 224; posterior 46, 49, 50, 58, 214–15; prior 13, 46–50, 57–8, 168, 214–15
problem of evil 11–12, 107–8, 114–24, 129, 134, 140, 143–5, 149–54, 157–8, 179–80, 184–9, 197, 204–5. *See also* atheism, arguments for; evil; suffering
problem of no best/unsurpassable world 8, 10, 132, 136–9
problem of suboptimality 149–59

quantum theory/quantum mechanics 3, 11, 27, 30–1, 34, 36, 41, 45–50, 54–8, 66–79, 85–7, 107, 109, 194

redemption 18, 224, 232, 236
resurrection 57, 89, 227, 232, 236

simplicity 11, 16–17, 45–7, 49–50, 52, 55, 57, 65–6, 168, 204, 229–30
sin 216, 227
states of affairs 16, 58, 87, 100, 104, 106, 118, 120, 132, 136, 139, 151, 159, 179, 181–9
string theory 3, 26, 34, 36, 39, 45, 50, 57–8, 62, 64, 85, 192, 211
suffering 16, 45, 51–7, 81–2, 108, 115, 123–4, 131, 141, 144–6, 154, 156, 159, 185, 188, 202, 219, 224, 227, 231; *see also* atheism, arguments for; evil; problem of evil
Swinburne, Richard 52, 67, 168, 180, 217–18, 220, 224–5

Tegmark, Max 3, 36, 39, 107–8
theodicy 38, 81–2
thermodynamics 29, 30
threshold, axiological 9, 10, 17, 123–4, 130–1, 137, 139, 231
Trinity 221, 223–5, 227, 232, 239

Van Inwagen, Peter 17, 67, 147, 159, 240

Zeno sphere 166, 167, 171